THE BASICS *of*
GEOMORPHOLOGY

KEY CONCEPTS

'Although there are many introductory geomorphology books, this latest contribution from Professors Gregory and Lewin is an excellent addition to the genre. They ask fundamental questions on the nature of scientific enquiry in geomorphology, on the burgeoning field of complexity studies and its interface with traditional geomorphological questions, and on the ways in which geomorphic processes relate to real world problems. The chapter on the Anthropocene illustrates well the variety of perspectives that geomorphologists have on this latest theme in geomorphology. The book is replete with boxes, web sites and recommendations for further reading. I would have appreciated having a copy of this book when I taught the introductory and follow-up geomorphology courses at the University of British Columbia.'
Olav Slaymaker, Professor Emeritus, Geography, The University of British Columbia and Former President of the International Association of Geomorphologists

'As Geomorphology has matured as a discipline, so it has developed a range of concepts which are fundamental for understanding it. This book, written by two leading geomorphologists, provides a very welcome first attempt to explain the major concepts in a concise and accessible way.'
Andrew Goudie, Emeritus Professor of Geography, University of Oxford

'Geomorphology and Earth sciences in general have been profoundly transformed by ideas that expand our vision from "normal" or "equilibrium" forms and developmental pathways (with exceptions to the pattern treated as deviations) to a much richer vision. The latter recognizes multiple possible pathways and outcomes, and landscapes and landforms that encompass, but transcend, classical equilibrium notions. *The Basics of Geomorphology: Key Concepts* is to my knowledge the first and only geography or geology textbook that fully embraces this new vision. Gregory and Lewin have managed the difficult but important task of distilling a new set of "basics" from these revolutionary insights that both broadens and complements traditional ideas. This book is remarkable in the way that it acknowledges a wonderful plurality of conceptual frameworks and approaches, and explains them in a clear, engaging way. Time will tell, but this book may well mark a turning point in the way students and scientists alike perceive Earth surface processes and landforms.'
Jonathan Phillips, University of Kentucky

'This book provides a very accessible overview of the discipline of geomorphology that is well supported by web-based material. The text is up-to-date, with excellent reference to relevant literature, and presents the discipline in an interesting and novel way. It provides a very readable and informative introduction to the discipline for senior undergraduates, postgraduates and researchers approaching the subject from other areas of science.'
Angela Gurnell, Queen Mary University of London

'In the current millennium humans have reached a critical point for coping with the surface of their planet, including its changing climate and environments, the alarming vulnerability to its hazards, and the increasingly complex problems of sustainability. All of these interrelated concerns require renewed attention to geomorphology, the science that deals with Earth's land surface and the processes that operate thereupon. I can think of no better guides than Professors Ken Gregory and John Lewin to lead the reader through the conceptual basis of this exciting science, as they do in their new book: *The Basics of Geomorphology: Key Concepts*. Geomorphology is a science that is being rapidly transformed by revolutionary new technologies for imaging, dating, and modeling Earth's dynamic surface. I highly recommend this book to those wishing for an up-to-date introductory overview of this rapidly changing science that is so critical to preserving the continuing habitability of our planet.'
Victor R. Baker, Regents' Professor of Hydrology and Water Resources (and Professor of Planetary Sciences and Geosciences), the University of Arizona

'Geomorphologists often struggle to understand and see the relevance of philosophy. This book provides an ideal introduction to geomorphological concepts packaged in a user-friendly and non-threatening way'.
Heather Viles, University of Oxford

'At a time when Earth's surface is changing rapidly, a geomorphology text centering on systems concepts, including interactions and adjustments, is especially timely. While grounded in key concepts of the discipline, this text prepares students for confronting new challenges ahead, including increasing human interactions in the "Anthropocene."'
Anne Chin, Department of Geography and Environmental Sciences, University of Colorado Denver

THE BASICS *of*
GEOMORPHOLOGY

KEY CONCEPTS

Kenneth J Gregory & John Lewin

Los Angeles | London | New Delhi
Singapore | Washington DC

Los Angeles | London | New Delhi
Singapore | Washington DC

SAGE Publications Ltd
1 Oliver's Yard
55 City Road
London EC1Y 1SP

SAGE Publications Inc.
2455 Teller Road
Thousand Oaks, California 91320

SAGE Publications India Pvt Ltd
B 1/I 1 Mohan Cooperative Industrial Area
Mathura Road
New Delhi 110 044

SAGE Publications Asia-Pacific Pte Ltd
3 Church Street
#10-04 Samsung Hub
Singapore 049483

© Kenneth J. Gregory and John Lewin 2014

First published 2014

Editor: Robert Rojek
Assistant editor: Keri Dickens
Production editor: Katherine Haw
Copyeditor: Audrey Scriven
Marketing manager: Michael Ainsley
Cover design: Francis Kenney
Typeset by: C&M Digitals (P) Ltd, Chennai, India
Printed in India at the Replika Press Pvt Ltd

Library of Congress Control Number: 2014933582

British Library Cataloguing in Publication data

A catalogue record for this book is available from
the British Library

MIX
Paper from
responsible sources
FSC
www.fsc.org   FSC® C016779

ISBN 978-1-4739-0574-0
ISBN 978-1-4739-0575-7 (pbk)

At SAGE we take sustainability seriously. Most of our products are printed in the UK using FSC papers and boards.
When we print overseas we ensure sustainable papers are used as measured by the Egmont grading system.
We undertake an annual audit to monitor our sustainability.

CONTENTS

ABOUT THE AUTHORS

Ken Gregory obtained his BSc, PhD and DSc from the University of London, was made CBE in 2007 for services to geography and higher education, and is currently President of the British Society for Geomorphology. He is Visiting Professor of Geography and Environment, University of Southampton. Research interests include river channel change and management, palaeohydrology and the development of physical geography, and he has written more than 140 papers, and authored and edited 30 books including *The Earth's Land Surface* (2010) and *The SAGE Handbook of Geomorphology* (2011). He has three Honorary degrees, and received the Founder's Medal of the Royal Geographical Society (1993), the Linton award of the BGRG (1999), and the Geographical medal of the Royal Scottish Geographical Society (2000).

John Lewin graduated with a BA and PhD from the University of Southampton. He was Professor of Physical Geography at Aberystwyth University, where he also served as Dean of the Science Faculty and Pro-Vice-Chancellor. His major research interest lies in fluvial geomorphology, especially floodplains, and the development of landforms over a full range of timescales. He is a former chairman of the British Society for Geomorphology and received the Society's Linton Award in 2011.

PREFACE

A headline in *The Times* for 17th May 2013 was '"*Medieval*" *lectures could be replaced by free online courses*'. This was stimulated by a conference hosted by UUK, called to consider the popularity of free online courses offered by many of America's leading universities. Views espoused included: students don't look forward to listening to a distant lecturer in a lecture hall with as many as 400 other students; students place high value on quality programme time, interactive content time with the academic community; lectures were medieval concepts introduced in Europe nearly 800 years ago and haven't changed much in that time; universities will be forced to rethink what they offer, with fees as high as £9,000 per year, as millions of people are now able to sign up for free courses anywhere in the world; new university buildings no longer include large lecture theatres; new methods of learning include providing 20 minute video lectures on line before the students attend timetabled classes with further discussion; flip lectures have developed where students watch a video or talk beforehand and then have questions or discussion in the lecture hall. These views illustrate dramatically the new environment in higher education learning, but how does it affect individual disciplines and books written for students? Perhaps no longer do they depend upon the comprehensive and often very substantial volumes of the past, together with separate treatments of a discipline, its philosophy and epistemology. In higher education do we now need books that focus on the very heart of the discipline, thus providing students with an anchor from which they can work? A foundation for independent, self-paced learning is required for this new and unfolding learning environment.

Concepts provide just such an integrating theme. In any discipline it is necessary to know those that have prevailed and evolved during the development of that discipline in order to understand the ones that are paramount in contemporary study. With the enormous volume of literature now available, much being too technically demanding to be understood easily, some students may be satisfied by knowing the basics about concepts. However, we hope that many will also want to know how the concepts were derived and to understand debates of the past and present. Students may be confused by the vigour and tenor of some debates, by those geomorphologists who take such a pessimistic view that it is difficult to discern what is left! In attempting to present a balanced view we see this book on at least two levels: first the basic content of the text, supported secondly by a web resource which enables enquiries to be updated, pursued further and in greater depth – exactly what academic enquiry is all about. Where the mass lecture is no longer the central element in higher education learning, the focus of this book can provide the basis for interactive questioning by students as they approach the discipline through enquiries related to the fundamental concepts. In that way this book provides access to the basics of landform science in a way appropriate for new learning environments.

This book aspires to provide a pedagogic text that gives a detailed description and analysis of key concepts for geomorphology. The intended readership encompasses senior undergraduates and early stage postgraduates looking for a full treatment of key concepts – where they come from and what they involve. It deliberately differs from textbooks by providing an in-depth treatment of the concepts, ideas and hypotheses that lie at the heart of geomorphology, rather than supplying a comprehensive region-by-region or system-by-system treatment of earth surface processes and forms. In order to present sufficient material in a short book so that it is central to student learning, we use tables to consolidate information, the web support to provide greater elaboration, together with extensive references and examples in *Progress in Physical Geography*. Topics for enquiry are suggested not only to illuminate concepts but also to facilitate geomorphological understanding. Now that recent conceptual developments have been absorbed it is very timely to focus on the underlying concepts, which are extremely appropriate for adaptive learning. This potentially enables what Bill Gates has identified as 'a special time in education'.

Acknowledgements

Every effort has been made to establish copyright, make acknowledgement and obtain permissions for figures, but if any have been missed please let us know.

Ken Gregory, John Lewin

HOW TO USE THE BOOK
AND WEBSITE

How to use this book

The Basics of Geomorphology: Key Concepts aspires to provide a pedagogic text that gives a detailed description and analysis of key concepts
for geomorphology, providing an in-depth treatment of the concepts, ideas
and hypotheses that lie at the heart of geomorphology, rather than a comprehensive region-by-region or system-by-system treatment of earth surface processes and forms. In order to present sufficient material in a short
book so that it is central to student learning, we use tables to consolidate
information, the web support to provide greater elaboration, together with
extensive references and examples in *Progress in Physical Geography*. We
want to provide access to the basics of landform science in a way that is
appropriate for new learning environments.

The book is complete so that a reader can gain sufficient understanding
of the background to geomorphology and its development (Chapter 1), to
major concepts which are each the subject of chapters (Chapters 2–19),
and then of geomorphology itself as a concept (Chapter 20). To assist the
reader a synopsis of content is given at the beginning of each chapter, and
references for further reading are given at the end of the chapter. Topics
are suggested to encourage further thinking about the subject matter of
the chapter or to be a focus for discussion. Citations in the text of Figures,
Tables and Boxes that are provided on the website are in bold so that
Figure X.n, **Table Y.n** or **Box.Z.n** will appear in this form.

Where the mass lecture is no longer the central element in higher education learning, the focus of this book can provide the basis for interactive
questioning by students as they approach the discipline through enquiries
related to the fundamental concepts.

How to use the website

The website (**study.sagepub.com/gregoryandlewin**) provides a resource
which:

- **gives detailed material** – to provide a more in-depth understanding
  of the content of the chapters, and thus several detailed tables have

been compiled which show how concepts developed; several chapters refer to **Boxes** which contain material that amplifies sections in the chapters of the book;

- **makes figures and diagrams available** – those that are essential for understanding the text are included in the book but those that elaborate points made or examples can be accessed on the website. Citations in the text of Figures, Tables and Boxes that are provided on the website are emboldened as **Figure X.n, Table Y.n or Box Z.n.**

- **provides a reference list** which includes all the references cited in the text: for a book of this kind the references cited are necessarily numerous, many students may not need to refer to all of them, but they can be accessed as required;

- **makes a glossary available** – throughout the book key concepts are the basis for book chapters and other concepts are mentioned (often shown in **bold** in the text) including many that have been used in other disciplines but are pertinent to geomorphology. The glossary has been compiled to aid readers by giving brief definitions for major concepts; it is not intended to be exhaustive because a glossary is available in *The Earth's Land Surface* (Gregory, 2010) and many definitions are provided in *The Dictionary of Physical Geography* (Thomas and Goudie, 2000); some terms have been adopted from the variety of parallel sciences but may be used in slightly different senses in geomorphology;

- **suggests relevant articles in *Progress in Physical Geography*** – there are many recent articles, including *Progress Reports*, that provide helpful up-to-date surveys of recent developments. By accessing these readers can be notified about the most recent developments.

The basic content of the text, supported by the web resource, enables enquiries to be pursued further, in greater depth and up-dated – exactly what academic enquiry is all about.

1

INTRODUCTION: CONCEPTS AND GEOMORPHOLOGY

Any discipline has concepts that are key for its progress. For geomorphology these need to be explicitly stated and consistently understood for what they are. We set them within the evolving history of geomorphology and the changing techniques and understanding that have been involved. This demonstrates the fashioning surges that have characterized the discipline and which complicate identification of those concepts which endure and which remain basic to the present and future study of geomorphology.

Geomorphology is the science concerned with the forms and processes on the Earth's land surface. Even with the fascinating challenges of explaining large-scale features such as the distribution of major mountain belts, spectacular landforms such as the Grand Canyon in the southwestern USA (**Figure 1.1**), or process events such as the Iceland volcanic eruption of Eyjafjallajökullin in 2010, geomorphology more generally has not captured the imagination of the general public as it could and should have done. There is today a need to advance geomorphic literacy in the way that climate literacy is being proposed (Climate Literacy, *Global Change Program 2009)*, to some extent redressing a lack of student knowledge (Theissen, 2011). It is arguable that practitioners of geomorphology have not sufficiently emphasized the interesting and important underpinning concepts of the discipline as much as they might have. Nonetheless, recent heightened awareness of global environmental change, including the effects of both rapid climate change and increasing human impacts on the Earth's surface, underlines the need to do so. A particular merit of geomorphology is that it has long recognized that our world is a changing one, and so the discipline has devised the means for interpreting sets of phenomena in terms of how they have developed in the past, and may develop in the future, beyond the timescales of individual human experience.

Geomorphology deals mostly with landforms on a human scale, although it may at the same time depend on comprehending physical, chemical and biological processes that range in their dimensions from the continental to the microscopic; forms range in size from minor slope features up to continent-scale landmasses. It also deals in timescales ranging from wind or water turbulence occurring in seconds to slow geological processes operating over millions of years: this poses rather obvious observational difficulties. There is a need to visualize happenings beyond everyday sensory appreciation, though no more so than for modern cosmologists studying the form and origins of the universe, or microbiologists deciphering the functioning of DNA. Landforms are also developed in complex spatial patterns that may be difficult to recognize at first sight. Over recent years many sciences have benefitted from the vastly greater availability of observational and analytical technologies. For geomorphology, this allows such things as the dating of Earth materials, the rapid assessment of sediment chemistry, the survey of river flows, computer modelling of emergent forms, and remote sensing of the Earth's surface. Key ideas can now be rigorously subjected to what in the business world are called 'proof of concept' (POC) procedures, using prototype field studies, laboratory analysis, and physical or numerical modelling; much contemporary published research is of this kind. Necessarily highly technical, this may initially prove alarmingly incomprehensible for many readers. But underlying it all, whether in the minds of researchers to start with or developed as observations proceed, there are key ideas that have come to crystallize our present understanding of the world's land surface.

The challenge for students of geomorphology, indeed of any discipline, is to find a comprehensible way into, and to become conversant with, modern research and its antecedents. With the advent of the internet and improved access to many sources, the research literature has burgeoned rapidly (with suggestions that the total amount of knowledge now doubles every 18 months), so it is easy to become overwhelmed with information and baffled by detail. In 2013 alone, the journal *Geomorphology* published 369 research papers, most of a highly technical nature. Consequently, many students have difficulty in distinguishing between basic underpinning concepts and useful but essentially research level technical material. This is not to belittle what are now indispensable techniques, or their on-going development that forms the focus for much dedicated present research. But for newcomers to the field without much technical knowledge, these can form an impenetrable initial barrier to understanding the things that the science is trying to do. One way of assisting such understanding is first to focus on key *concepts*, which can be defined as **those abstract ideas, general notions or units of knowledge that are vital to the development of a reliable science.** There is a

series of connections (Harvey, 1969: 19) between sense perceptions (per-
cepts), mental constructs and images (concepts) to linguistic representations (terms). Although hitherto concepts have not featured prominently
in any one geomorphology book it is important that we consider which
concepts actually do underpin geomorphological thought (major concepts are shown in **bold** in subsequent chapters), and how such abstract
or general ideas have been deduced or inferred from specific empirical
data. Knowledge of these concepts, and their development, can provide
the gateway to a more general understanding of what landform science
is currently able to tell us.

Philosophy is the discipline concerned with an a priori analysis of
concepts, as ideas are sought, possessed or understood, in coming to
formulate beliefs (and ultimately knowledge) about the real world.
Philosophers have given considerable attention to concepts, since
Immanuel Kant (1724–1804) characterized those resulting from experience as *a posteriori*, and Arthur Schopenhauer (1788–1860) contended
that concepts are 'mere abstractions from what is known through intuitive perception'. Philosophers and others have recognized several types
of concept although Machery (2009) argued that the dominant psycho-
logical theories of concepts have yet to be organized within a coherent framework. Laurence and Margolis (1999) suggested that there is
still much controversy about what kinds of things concepts are, how
they are structured and how they are acquired. For many, one of the
traditional tasks of analytic philosophy is that of providing analyses of
concepts which can be thought of as mental representations, as abilities
peculiar to cognitive agents, or as abstract objects. *Frames*, which originated in logic-based artificial intelligence (AI), have been suggested as
the basic format for concept formation in cognition because they are an
excellent tool for the investigation of conceptual frameworks underlying
scientific *theories* (hypotheses related by logical or mathematical arguments explaining a variety of connected phenomena) and their respective *ontologies* (the set of entities presupposed by theories) (Schurz and
Votsis, 2007). A dynamic frames approach was developed by Lawrence
Barsalou (1992) for the representation of concepts and the addressing
of issues about conceptual change (see Andersen and Nersessian, 2000).
Thus *Key Concepts in Geography* (Holloway et al., 2003) listed concepts
including space, time, place, scale, physical systems, landscape and environment. Subsequently in a second edition Clifford et al. (2009) added
nature, globalization, development, sustainability and physical geography, and risk. The breadth of things that could be regarded as underpinning concepts, explicit or implicit, is extremely wide.

For this book, we focus on concepts necessary for our current understanding of geomorphology as landform science. Those selected are less
general than time and space, but also less specific than individual logical

or mathematical theories linking entities (such as the relationship between river channel dimensions and discharge variables). Other expressions used for over-arching approaches are *paradigms* (Kuhn, 1962) and *research programmes* (Lakatos, 1970), but here we focus on concepts that relate specifically to geomorphology. In the remainder of this introduction we outline the development of geomorphology as a discipline, and its relationship to physical geography and to geology, ecology and environmental science (1.1). Subsequently we indicate the broad categories of techniques employed by geomorphologists (1.2), and then explain the chapter structure for the presentation of concepts in four major sections, indicating why the specified concepts have been selected (1.3).

1.1 Geomorphology as a discipline

In addition to four books in a series on the history of the study of landforms (Chorley et al., 1964; 1973; Beckinsale and Chorley, 1991; Burt et al., 2008) other works have traced the antecedents of the science of geomorphology in detail so that only a synopsis of the development of the science is provided here. The first use of the word geomorphology was in 1858 in the German literature (see Roglic, 1972; Tinkler, 1985). Table 1.1 suggests some of the founding assumptions prior to that date and indicates others that were subsequently influential. In geomorphology, as in any other discipline, the foundations of the discipline are significantly shaped by the contributions of single individuals, but similar ideas can emerge in more than one country. Whereas in the 19th and early 20th centuries ideas diffused relatively slowly, by the late 20th and early 21st centuries the speed of technical communication was so rapid that it is no longer easy to identify a single influence or the origin of ideas. Hence Table 1.1 is an approximation, compiled as a series of indicator milestones, to suggest the influences affecting the shaping of a discipline from the mid 19th century to the present day. Many ideas are cross disciplinary so that terms such as 'evolution', which would have been without great scientific meaning before about 1800, are now used (though with different technical definition and process underpinnings) from astrophysics to biology. In addition to individuals, there are wider influences: broader scientific ideas, researchers' experience of particular regions, diverse publications and journals, the work of academic societies and technological developments, as well as the constraints set by financial limitations or the public policy demands of particular societies. The selection in Table 1.1 tells much about our perception but it can be supplemented by the reader. A different example reflecting writer perception is the cast of principal characters listed by Kennedy (2006) which omits some that we include but includes others that we do not.

In the light of developments summarized in **Box 1.1** three particular motivational trends now recognized are: the need for more multidisciplinary research and investigations; the question about how far geomorphology can extend; and the potential to make further progress in relation to the management and design of environments (Gregory and Goudie, 2011b). However, a paradox now appears: despite the obvious importance of, and interest in, geomorphology as concerned with explaining the land surface of the Earth, the discipline itself 'remains little known and little understood, certainly in relation to other academic disciplines, and especially outside university circles' (Tooth, 2009). Appreciating the way that the science of landforms has grown as a discipline, and being mindful of the potential that it now has, mean that this is a particularly appropriate time to focus upon basic fundamental concepts. **Table 1.1** provides the background for such concepts against the timeline of the development of geomorphology.

1.2 Techniques employed by geomorphologists

Technical advances are arguably enabling great progress to be made in the 21st century. In fact since 1960 changes in geomorphology are reminiscent of the way in which Chemistry and Physics were changed by the technological breakthroughs in the early 20th century. Thus Summerfield (2005b) described the subject as having 'major research frontiers ranging in scale from the transport paths of individual particles over a river bed to the combined tectonic and surface processes responsible for the 100 million year history of sub-continental scale landscapes'. Before the 1960s geomorphology paid little explicit attention to techniques but used time-consuming methods of field surveying, field sketching, and mapping as appropriate. The way in which the scope of techniques available for the geomorphologist has changed is demonstrated by the content of books written to summarize those available (Table 1.2); to demonstrate the consequences of this explosion of techniques, Table 1.3 presents six categories, with an indication of those available fifty years ago, key developments since that time, and an outline of the contemporary range. **Table 1.4** summarizes the range of dating techniques now available.

Many examples could be given of recent dramatic progress made possible, but DEMs of difference (DoDs), which quantify volumetric change between successive topographic surveys, and Structure for Motion (SfM) methods, which estimate three-dimensional structures from two-dimensional image sequences, illustrate the progress now possible. Overviews from space both allow geomorphologists not only to see and measure the characteristics of large landforms, but also to

Table 1.2 Examples of publications reflecting the development of techniques in geomorphology

Year	Publication	Major Contents
1966	*Techniques in Geomorphology* C.A.M. King	Observation of form and character; Observation of processes in action; Experiment and Theory (models); Cartographic and morphometric analysis; Sediment analysis and statistical analysis
1969-	*Technical Bulletins* British Geomorphological Research Group	Aim was to 'have a source of standardized information relating to increasingly sophisticated methods of data collection' 26 published by 1980
1981	*Geomorphological Techniques* A. Goudie	In five parts: 1. Introduction, 2. Form, 3. Material properties, 4. Process, 5. Evolution
1990	*Geomorphological Techniques* A. Goudie (2nd edition)	
1983	*Geomorphological Field Manual* R. Dackombe and V. Gardiner	Chapters: 1. Topographic survey, 2. Geomorphological mapping, 3. Slope profiling, 4. Mapping landscape materials, 5. Geophysical methods of subsurface investigation, 6. The description of landscape forming materials, 7. Fluvial processes, 8. Glacial processes, 9. Aeolian processes, 10. Coastal processes, 11. Slope processes, 12. Sampling, 13. Miscellaneous aids
2003	*Tools in Fluvial Geomorphology* M. Kondolf and H. Piégay	21 chapters in seven sections: I. Background, II. The Temporal Framework, III. The Spatial Framework, IV. Chemical, Physical and Biological Evidence: Dating, V. Analysis of process and forms, VI. Discriminating: simulating and modelling processes and trends, VII. Conclusion: Applying the tools
2010	*Geomorphological Techniques* (Online Edition) British Society for Geomorphology ISSN 2047-0371	Organized in five sections: Composition of Earth Materials Topographic and Spatial Analysis Processes, Forms and Materials in Specific Environments Long-term Environmental Change (dating techniques, etc.) Modelling Geomorphic Systems

Table 1.3 A view of techniques for geomorphology

Methods, Tools and Techniques	1960s	Subsequent Key Developments
Field techniques	Field mapping, some examination of sediments and deposits, few process measurements	Electronic distance measurement (EDM) Global positioning systems (GPS) Geographical information systems (GIS) automatic loggers Analysis in real time Remote access recording of continuous measurements Geomorphological mapping revived
Numerical techniques	Pre-high speed electronic computers, very basic mathematical and statistical models	Quantitative and statistical analysis Focus on the general rather than the individual characteristics of a particular area Progress from linear to non-linear methods, chaos, and nonlinear dynamical systems approaches, GIS
Remote sensing	Air photographs	Remote sensing providing frequently repeated imagery Sensing of many aspects not previously possible LiDAR gives major advances Prospects of global DEMs at better than 10 m resolution and complete image coverage of the Earth at better than 50 cm Google Earth
Laboratory analysis	Limited to time consuming chemical analysis and size analysis of deposits by sieving and titration	Laboratory analysis of samples of water and sediments rapidly achieved Great range of new instruments available for analysis of rock, sediment, and fluid samples
Modelling	Qualitative models of long-term landscape development	Advances in numerical modelling, geochronology and remote sensing quantitative techniques used to model earth surface process and interpret the landscape
Dating techniques	Some ^{14}C	Great range of new dating methods, some very innovative Quaternary chronology refined Cosmogenic dating enabled great advances in deducing rates of erosion

focus on metre scale forms in remote environments. Geomorphology is becoming far less concentrated on those landforms which are easily accessible in mid-latitude long-populated environments.

1.3 A structure for concepts

Many concepts in geomorphology have been generated following the empirical study of specific instances or occurrences of phenomena. These can be used to encapsulate the commonality between seemingly disparate localised phenomena, so that one would expect geomorphological texts to be concerned with linking concepts as basic to the discipline. Table 1.5 collates works that cite concepts but it is striking that relatively few are explicitly conceptual in format; geomorphological research appears to have been empirically rather than theoretically driven. What Table 1.5 does show is the implicit conceptual guidelines – the ideas, assertions and hypotheses that have accumulated and evolved during the development of the discipline. We are well aware that many concepts could be considered as 'key' for the discipline in detail, but we believe our selection is appropriate for understanding the overall nature, and the attraction and challenge, of modern geomorphology. Complementary information on geomorphological concepts is provided through texts of several kinds (Table 1.6).

Recent years have seen the transformation of geomorphology with the advent of substantial advances in techniques available, but unlike the physical sciences the provision of general laws may be considered an unrealistic dream in view of the contingent factors that are people-, place- or time-dependent. It has been suggested that this is as important or even more important than general laws in determining how the world works (Phillips, 2004). Phillips (2004) therefore sees a future requirement as confronting the creative tension between *nomothetic* (concerned with general theories) and *idiographic/interpretive* (concerned with individual cases) science, and integrating the two approaches.

So what is the most appropriate way of organizing geomorphological concepts for the discipline in the 21st century? Any structure has to reflect the way in which geomorphology has evolved, it has to be capable of embracing the range of concepts which are now the basis for geomorphological study, and it should form an appropriate platform for taking on the challenges of the 21st century. The first group of concepts (A in Table 1.7) focuses on system contexts involving the methodology of the discipline and the way in which investigations are approached. Systems have provided the most durable conceptual approach since promoted by Chorley (1962) fifty years ago, so that the three subsequent categories are concerned with the functions

Table 1.7   Book Structure

___

1. Introduction: Concepts and Geomorphology

SECTION A: System Contexts

2. The Systems Approach
3. Uniformitarianism
4. Landform
5. Form, Process and Materials
6. Equilibrium
7. Complexity and Non-linear Dynamical Systems

SECTION B: System Functioning

8. Cycles
9. Force-Resistance
10. Geomorphic Work
11. Process-form Models

SECTION C: System Adjustments

12. Timescales
13. Forcings
14. Change Trajectories
15. Inheritance
16. The Anthropocene

SECTION D: Drivers for the Future

17. Geomorphic Hazards
18. Geomorphic Engineering
19. Prediction and Design

CONCLUSION

20. The Concept of Geomorphology

___

(B in Table 1.7), adjustments (C), and present and future management (D) of the geomorphic system. These four categories provide a logical sequence which, when explored, can prove challenging and thought-provoking. In subsequent chapters, as in this one, tables are employed, both within the text and available online, to give additional detailed information. These are not essential to an understanding of the text but provide more detailed background information that the enquiring reader may require.

FURTHER READING

Gregory, K.J. and Goudie, A.S. (2011) Introduction to the discipline of geomorphology (pp. 1–20), Conclusion (pp. 577–85). In K.J. Gregory and A.S. Goudie (eds), *The SAGE Handbook of Geomorphology*. London: Sage.

Murray, B., Lazarus, E., Ashton, A., Baas, A., Coco, G., Coulthard, T., Fondstad, M., Haff, P., McNamara, D., Paola, C., Pelletier, J. and Rheinhardt, L. (2009) Geomorphology, complexity, and the emerging science of the Earth's surface, *Geomorphology*, 103: 496–505.

Phillips, J.D. (2012) Storytelling in the Earth sciences: the eight basic plots, *Earth Science Reviews*, 115: 153–62.

Tooth, S. (2009) Invisible geomorphology, *Earth Surface Processes and Landforms*, 34: 752–54.

TOPICS

1. In the light of the philosophy background to concepts, the history of the development of geomorphology and the list suggested in Table 1.7, where would you position concepts of evolution, and what other concepts would you expect to be included?

WEBSITE

For this chapter the accompanying website **study.sagepub.com/gregory andlewin** includes Figure 1.1; Tables 1.1, 1.4, 1.5, 1.6; Box 1.1; and useful articles in *Progress in Physical Geography*. References for this chapter are included in the reference list on the website.

SECTION A

SYSTEM CONTEXTS

2

THE SYSTEMS APPROACH

A system as a set of components and relationships between them, function-ing to act as a whole, has been detectable in science and in thinking about landforms for more than a century. For geomorphology, it was formalized in 1962 when the benefits of an open systems approach were articulated. The approach has become integral to many aspects of landform science, has been accompanied by other conceptual developments, and has been suc-ceeded by self-organizing systems with non-linear relationships and more uncertainty.

The idea of a system is not new: Newton wrote of the solar system, biologists have been concerned with living systems, geographers have implicitly used the systems concept since the early days of the subject (Gregory, 2000), and most Earth scientists have probably always thought in systems terms. In Hutton's rock cycle, from his *Theory of the Earth* (Hutton, 1788, 1795), the Earth is, in effect, being described as a sys-tem, with different system components, materials and processes through which matter is transported and recycled (Odoni and Lane, 2011). The system is now frequently employed in many scientific disciplines and, to give another example, Lovelock (2009) used Webster's *New Collegiate Dictionary* definition: 'an assemblage of objects united by some form of regular interaction or interdependence'. Geomorphological systems may now be viewed as one component of an immensely complex total Earth System including atmospheric, oceanic, biological and other elements.

As with all concepts, it is not easy to discern exactly how the **systems concept** originated and when it was first applied. In addition to early ideas about Earth systems a late 19th and early 20th century trend in the physical, especially chemistry and biological, sciences was towards a recognition of systems, and it was probably ideas from biology that led Ludwig von Bertalanffy (1901–1972) to propose *General systems theory* as an analytical framework and procedure for all sciences. Much of his published work in the field of 'organismic' biology was written in German and is thus not widely known (Drack, 2009) so that from

1932 it was not immediately absorbed elsewhere. There were several antecedents such as von Uexküll (**Table 2.1**) that set the stage for von Bertalanffy's 1937 *General systems theory* proposal at a philosophical seminar in Chicago and formalizing systems theory in 1950.

After outlining the adoption in geomorphology (2.1), we review the implications of systems approaches embedded in landform science (2.2), to see how this is continuing to evolve (2.3).

2.1 Adoption of the systems approach in landform science and geomorphology

Strands of ideas in geomorphology anticipated the advent of systems theory. These included those from G.K. Gilbert (1843–1918) in 1877, from J.T. Hack in 1960, and from A.N. Strahler (1918–2002) in 1952, but we could also add J.F. Nye's (1952) application of plasticity theory to the flow of ice sheets and glaciers, and the classic work of R.A. Bagnold (1896–1990) on the physics of blown sand and desert dunes (Bagnold, 1941). Such strands (**Table 2.1**) provided the context, but it was the classic paper by R.J. Chorley (1927–2002) in 1962 that really embedded systems thinking in geomorphology. He contrasted the open with the closed system view that was at least partly embodied in Davis's view of landscape development. Whereas open systems require an energy supply for maintenance and preservation, maintained in an equilibrium condition by the constant supply and removal of material and energy, in a closed system the given amount of initial free energy becomes less readily available as the system develops towards a state with maximum *entropy*, signifying the degree to which energy has become unable to perform work. The value of the open system approach to geomorphology was summarized as having several useful purposes (Chorley, 1962: B8):

- To show dependency on a universal tendency towards an adjustment of form and process.
- To direct investigation towards the essentially multivariate character of geomorphic phenomena.
- To admit a more liberal view of morphological changes with time, to include the possibility of non-significant or non-progressive changes of certain aspects of landscape form through time.
- To foster a dynamic approach to geomorphology to complement the historical one.
- To focus upon the whole landscape assemblage rather than those parts assumed to have evolutionary significance.

- To encourage geomorphic investigations in those areas where the evidence for erosional history may be deficient.

- To direct attention to the heterogeneity of spatial organization.

However, Huggett (2007), in his review of the systems approach in geomorphology, suggested that it was Strahler (1950; 1952: see also 1980), rather earlier, who introduced open systems theory to geomorphology, ushering in a revival of Gilbertian thinking. This involved concepts drawn from physics and mechanics rather than historical geology. Exemplifying the difficulty of pinpointing the actual source of concepts, the papers by Strahler and by others (**Table 2.1**) certainly provided foundations for a different way of analysis which then progressed to explicit systems approaches. The four types of system recognized (morphological, cascading, process-response, control) by Chorley and Kennedy (1971), as well as the four phases distinguished (lexical, parsing, modelling, analysis) by Huggett (1980) and their subsequent adoption, are described in **Box 2.1**.

2.2 Embedding and encompassing the systems approach

Systems ideas have prompted – or at least combined with – other conceptual developments required for the further development of landform research.

Inclusion of the systems approach was important for modelling and provided a context for ideas such as equilibrium; was parallel with other ideas such as land systems and contributed to others such as Earth system science; was helpful in reconciling timeless and time bound approaches; and could stimulate new ideas and provide the basis for new developments.

Their vital importance is shown in many aspects of **modelling**. Odoni and Lane (2011) considered that a system can be imagined as having the properties of (1) objects (e.g., a grain of sediment); (2) processes that act on objects (e.g., momentum transfer whether from a fluid or other grains, that makes the grain move) and which connect objects together, and which are often specified in the form of rules or algorithms; (3) boundaries, often introduced to make the modelling problem tractable (e.g., defining the spatial extent of the deposit over which sediment movement will be simulated); (4) boundary conditions, necessary to recognize that when boundaries are involved, additional or auxiliary information is required (e.g., the sediment feed rate); and (5) exogenous drivers that cause change in the boundary conditions (e.g., a change in sediment feed rate). These five properties are essential for the structure of many models. Over the last four decades many geomorphic models have been structured upon a systems framework, which encourages modelling involving both forms and the transfer of energy and materials necessary to analyse dynamic changes in geomorphology.

Use in relation to other concepts included **equilibrium**: this is a concept which has been thought about for more than one hundred years, and has now been conceived of in several ways including steady state, dynamic, or metastable equilibrium (see Chapter 6). It was placed in context by the advent of systems thinking. An *open system* condition may be assumed in which quantities of stored energy or matter are adjusted so that the input, throughput and output of energy or matter are balanced. Although the conceptual frameworks of systems analysis and geomorphic equilibrium can be divergent in many respects (Mayer, 1992), it has been argued that many geomorphic system states and behaviours, often interpreted as showing a tendency towards the establishment and maintenance of steady-state equilibrium, are actually emergent outcomes of two simple principles – gradient selection and threshold-mediated modulation (Phillips, 2011b). It is explained in Chapter 6 that the contemporary interpretation of equilibrium is significant and useful but not universal, does not necessarily have a single final equilibrium state, and can be visualized in different ways including as a metaphor.

The **land systems approach** is an example of a specific development that was aided by the systems approach. Resource surveys introduced by the Australian Commonwealth Scientific Industrial Research Organization (CSIRO) in 1946, designated land systems as areas or groups of areas with recurring patterns of topography, soils and vegetation with a relatively uniform climate. The implementation of this approach, employed especially for the management of resources and modified for application to urban and suburban areas, was greatly advanced by satellite remote sensing and GIS. The approach has been adapted for other geomorphological studies of landscapes, especially those that include multiple remnant components making up an overprinted palimpsest of former conditions (see Chapter 15). Thus as the withdrawal of glacier ice exposed landscapes, usually over timescales of $10^1->10^4$ years, six paraglacial landsystems have been identified (Ballantyne, 2002a): rock slopes, drift-mantled slopes, glacier forelands, and alluvial, lacustrine and coastal systems – each containing a wide range of paraglacial landforms and sediment facies. **Paraglacial** (Church and Rider, 1972) may be defined as 'non-glacial Earth surface processes, sediment accumulations, landforms, land systems and landscapes that are directly conditioned by glaciation and deglaciation' (Ballantyne, 2002a; Ballantyne, 2003), so that the paraglacial is the period of readjustment from glacial to non glacial conditions (Church and

Slaymaker, 1989; Slaymaker and Kelly, 2007:167), and research data on rates of operation of some paraglacial systems have been compiled (Ballantyne, 2002a). Striking landforms in the area of the southern Laurentide ice sheet were analysed in terms of seven land systems (Colgan et al., 2003). The Satujökull foreland of the northern Hofsjökull ice cap in central Iceland shows a clear signature of glacial land system overprinting as a result of complex glacier behaviour during the historical period. Landsystem 1, comprising a wide arc of ice-cored moraine and controlled ridges lying outside fluted and drumlinized terrain, is indicative of polythermal conditions. Landsystem 2 contains most of the diagnostic criteria for a surging glacier landsystem with records of two separate surges. Observation of landsystem overprinting, especially in response to changing thermal regimes and/ or glacier dynamics, and particularly by different flow units in the same glacier, is rarely reported but is crucial to the critical application of modern landsystem analogues to Quaternary palaeo-glaciological reconstruction (Evans, 2011).

More recently developed is **Earth System Science.** Appearing in 1988 (NASA, 1988) and stated in the Amsterdam declaration of 2001, this is the study of Earth as a total system with various components, such as the atmosphere, hydrosphere, biosphere and lithosphere. It therefore embraces geomorphology. It has been suggested (Lovelock, 2009) that this concept grew within the Earth science community to form an intellectual environment for explaining the flood of new knowledge about the Earth. It arose from Gaia theory but did not encompass habitability as the goal for the self-regulation of the Earth's climate and chemistry. Although it is seen as an all-embracing science envelope, Clifford and Richards (2005) concluded that earth system science (ESS) constitutes an oxymoron; it should be seen neither as an alternative to the traditional scientific disciplines, nor regarded as a wholesale replacement for a traditional vision of environmental science, but rather as an adjunct approach. Subsequently it was suggested (Richards and Clifford, 2008) that LESS (local environmental systems science) would be a more appropriate focus for geomorphology. Perhaps **Gaia theory** – introduced by Lovelock in the 1980s as 'a view of the Earth that sees it as a self-regulating system made up from the totality of organisms, the surface rocks, the ocean and the atmosphere tightly coupled as an evolving system' (Lovelock, 2009) – is the best example of an idea that had developed from systems and provides a context for them. Lovelock (2009) quotes the Nobel Prizewinner Jacques Monod (1970) who drew attention to holistic schools which, phoenix-like, are reborn in every generation, and the analytical attitude (reductionist) was doomed to fail in its attempts to reduce the properties of a very complex organization to the 'sum of its parts'. However, a systems approach can accommodate both holistic and reductionist approaches.

The systems approach could reconcile **timeless** and **time bound** approaches. When the systems approach was developed in the 1960s it was associated with the surge of process geomorphology (timeless) and, at that stage, was almost independent from research undertaken on landscape development (time bound). However, systems can be the basis for a reunification of the two approaches, as exemplified by the use of land systems which may include inherited elements. The likelihood of any landscape or geomorphic system existing at a particular place and time with such effect as to exclude all its predecessors is negligibly small. This idea has also been extended in terms of the 'perfect landscape', conceptualized as being the result of the combined interacting effects of multiple environmental controls and forcings to produce an outcome that is highly improbable, in the sense of duplication at any other place or time. Geomorphic systems have multiple and variable environmental controls and forcings, which allow for many possible landscapes and system states (Phillips, 2006a). The analogy here is with 'the perfect storm' that arises when all possible formative factors occur together. A perfect landscape perspective (Phillips, 2007) leads toward a world view that landforms and landscapes are circumstantial, contingent results of deterministic laws operating in a specific environmental context such that multiple outcomes are possible (see Chapters 7 and 15). This contrasts with the earlier view embracing single outcomes for a given set of laws and initial conditions. Thus Huggett (2007) sees this as a powerful and integrative new view, proposing landscapes and landforms as circumstantial and contingent outcomes of deterministic laws operating in a specific environmental and historical context, with several outcomes possible for each set of processes and boundary conditions. If capable of reconciling different geomorphological traditions, this could be a great success for the systems approach.

The **development of new ideas** has been fostered by the adoption of the systems approach, as shown by some of the above examples. Some have developed within geomorphology as illustrated by the instability principle developed by Scheidegger (1983) to connote the way in which equilibrium in geomorphic systems is commonly unstable. Any deviation from the equilibrium state may be self-reinforcing, causing the deviation to grow with increasing irregularity over time, as illustrated by the growth of cirques. Linking ecological and geomorphological systems, often previously largely conceptualized as independent, has fostered other ideas. Ecological 'memory', which encompasses how a subset of abiotic and biotic components are selected and reproduced by recursive constraints on each other, is reflected in the way that on-going interactions between ecology and geomorphology become encoded in the landscape (Stallins, 2006). What this means is that repeated interactions make up a history or trajectory of change in which what follows at each stage is determined by the interactions which have taken place before, and not just the ones that can be observed at particular points in time. Disturbance regimes are a further example showing how non-linear systems react to human-induced and natural disturbances, illustrated by arid hillslopes, weathering systems in deglaciated areas and vegetated dunes in drylands (Viles et al., 2008). More widely, 'disturbance' can refer to any externally driven perturbation: geomorphology can be concerned with how landforms and landscapes respond to disturbances and to variable boundary conditions, and hence to how geomorphic systems co-evolve with climate, ecosystems, soils and other environmental systems (Phillips, 2011a). This allows the assessment of geomorphic changes and responses to be based on *response* (reaction and relaxation times), *resistance*, *resilience* (recovery ability) and *recursion* (positive and/or negative feedbacks) – the 4 Rs put forward by Phillips (2009; 2011). Small-World Networks (SWN) are networks with a special structure that model the relationships in the real world including those in ecosystems, so that it is possible to visualize geomorphic systems as coupled subsystems with SWN traits characterized by tightly connected clusters of components, with fewer connections between the clusters (Phillips, 2012a).

Thus Thornes and Ferguson (1981) extended conceived applications from *simple systems* (which involve no more than three or four variables, utilize Newtonian laws, and can be handled by relatively simple techniques, including regression models and partial differential equations possibly extending to finite difference methods) to *systems of complex disorder* (involving large numbers of components/variables but only weak linkages between them, requiring probabilistic methods of statistical mechanics, including probabilistic approaches to soil creep and to stream networks, coastal spit simulation and Box Jenkins models). Systems of complex order were also recognized and these have been developed in conjunction with non-linear dynamical systems in a complexity approach to the interpretation of landforms (see Chapter 7).

2.3 Conclusion

Systems are now an integral part of geomorphology. The concept originated in other sciences and especially biology, but now embedded in geomorphology it is fundamental in facilitating significant developments in the discipline, especially those associated with non-linear dynamical systems. Furthermore, as Chorley (1962) stated in his seminal paper, 'It is only through . . . application of systems analysis that considerations of the management of the natural environment can be elevated above mere *ad hoc* book-keeping to form part of a broader scholarly discipline which focuses on the conservational aspects of geographical control systems'. As geomorphology is increasingly concerned with the application of research results, perhaps this may provide the greatest justification for the systems approach. However, systems continue to be debated. Von Elverfeldt and Glade (2011) argued that the theoretical foundation as well as the definitions and basic assumptions are rarely (if at all) questioned, subsequently suggesting (von Elverfeldt, 2012) a view of systems as being open but at the same time operationally closed, as self-organized structure-building and potentially self-referential.

FURTHER READING

Chorley, R.J. and Kennedy, B.A. (1971) *Physical Geography: A Systems Approach.* London: Prentice Hall.

Gregory, K.J. (2000) *The Changing Nature of Physical Geography.* London: Arnold. (See especially Chapter 4.)

Huggett, R.J. (2007) A history of the systems approach in geomorphology, *Géomorphologie: Relief, Processus, Environnement,* 2: 145–58.

Phillips J.D. (1999) *Earth Surface Systems: Complexity, Order, and Scale.* Oxford: Blackwell.

von Elverfeldt, K. and Glade, T. (2011) Systems theory in geomorphology; a challenge, *Zeitschrift für Geomorphologie, Supplementary Issues,* 55: 87–108.

TOPICS

1. Consider the proposals in von Elverfeldt and Glade (2011) and von Elverfeldt (2012) – are they likely to advance systems underpinning of geomorphology or are they a distraction?

 ## WEBSITE

For this chapter the accompanying website **study.sagepub.com/ gregoryandlewin** includes Table 2.1; Box 2.1; and useful articles in *Progress in Physical Geography*. References for this chapter are included in the reference list on the website.

3

UNIFORMITARIANISM

*Often interpreted as 'the present is the key to the past', uniformitarian-
ism has been an important concept influencing the development of Earth
sciences since it was introduced in 1832. It is acknowledged as a stimulating paradigm influencing geomorphological thinking, but more recently
has been critically reviewed, considered to be integral throughout science,
and complemented by other concepts including actualism, gradualism and
catastrophism.*

Uniformitarianism is the concept often expressed as the 'present is the
key to the past' (Geikie, 1905). Explanations could be framed in terms
of observable processes rather than ascribed to catastrophic events or
the intervention of a deity. It is consistent with the *heuristic* (general
guiding rule based on experience or observation) which has become
known as Occam's (or Ockham's) razor, a principle attributed to the
14th century Franciscan friar William of Ockham, and which has often
been simplified as 'other things being equal, a simpler explanation is
better than a more complex one', or for scientists 'when you have two
competing theories that make exactly the same predictions, the simpler
one is the better'. 'Simplicity' in this context means interpreting phenomena in terms of observations or unexceptional processes rather than
events like the Biblical flood or primeval catastrophes.

However, this is not always straightforward. For example, the origin
of erratics caused a great deal of controversy in the mid 19th century,
notably between Charles Lyell (1789–1875) and Louis Agassiz (1807–
1873), with the former championing the idea that they had drifted
into position on icebergs floating in a once deeper ocean, and the latter arguing for direct transport on much-extended glaciers. As we shall
see below Lyell is regarded as one of the great champions of uniformitarianism, but neither Lyell nor Agassiz was able directly to observe the
agents they invoked – namely a deeper ocean or a vastly expanded ice
mass flowing over very gentle gradients. Lyell could indeed be said in

part to follow uniformitarian principles (he saw floating icebergs in the Atlantic, though he imagined the deeper ocean), but we now recognize that he was wrong. Many others in the 19th century didn't believe in an 'ice age' at all, holding that the history of the Earth was one of progressive cooling; they found an ice age simply 'inconceivable'.

In works that have specified concepts (**Table 1.5**), it is notable that virtually all include uniformitarianism explicitly. In other words, this is a basic concept in geomorphology – one that has been central to our understanding of the Earth's surface – but it has also been applied in different contexts, and it is necessary to appreciate how it has evolved (3.1), its relationship to alternative or additional concepts (3.2), and its present significance (3.3).

3.1 The development of uniformitarianism

Until the late 18th century prevailing schools of thought explaining the Earth's surface were teleological in that they viewed God as responsible, with a succession of catastrophes (including Noachian deluge) being divinely inspired. Strong views prevailed, although the notebooks of Leonardo da Vinci (1452–1519) show that he may have developed deductive methods to understand surface processes, and Pierre Perrault (1611–1680) showed that precipitation was sufficient to sustain the flow of the Seine river, in contrast to the long-held belief that subterranean condensation or the return flow of sea water was responsible for river discharge. Strong views of the early catastrophists included those of Dean Buckland who, following Archbishop Ussher (1581–1656), believed that the world began in 4004 BC. Views began to change with the *Theory of the Earth* published as two volumes by James Hutton (1726–1797) in 1795 and subsequently clarified by Playfair (1747–1819) in his *Illustrations of the Huttonian Theory of the Earth* (1802). This theory rejected catastrophic force as the explanation for environment and gave rise to a school of thinking whereby a continuing uniformity of existing processes was regarded as providing the key to an understanding of the history of the Earth. This replaced catastrophic ideas of landscape formation, with the idea promulgated that 'the present is the key to the past' and the doctrine of uniformitarianism attributed to Charles Lyell (1797–1875), who published his *Principles of Geology* in 1830, and who came to be regarded as 'the great high priest of uniformitarianism'. Ironically Lyell studied under the catastrophist William Buckland at Oxford, but became disenchanted with his tutor when Buckland tried to link catastrophism to the Bible. Although the idea behind uniformitarianism originated in this way, the term **uniformitarianism** was first used in 1832 by William Whewell (1794–1866). As summed up by Record (1997: xix):

'Lyell's recommendations in the *Principles* about method are turned into doctrines or 'isms', and compared with those of his contemporaries, the Cambridge polymath William Whewell began this process by dubbing Lyell a 'uniformitarian' in opposition to his own 'catastrophism'. Uniformitarianism was taken to mean that the long-continued operation of existing processes provides a framework for understanding the geomorphic and geologic history of the Earth. Geikie (1905: 299) expressed this idea as 'the present is the key to the past'.

It is now appreciated that uniformitarianism assumes that everything in nature can be explained 'scientifically'; it does not have to be attributed to the operational intervention of an over-riding deity. The assumptions involved may be that there is *uniformity of law* – the assumption that natural laws are constant in time and space; *uniformity of process* (**actualism**); *uniformity of rate* (**gradualism**); and *uniformity of state* (**steady statism**).

Although many geomorphologists (e.g., Ahnert, 1998) have said that James Hutton introduced uniformitarianism, in fact Hutton's ideas were not widely accepted by the scientific community until Lyell published his three volumes on *Principles of Geology* (Lyell, 1830–1833) in which he used a variety of evidence from England, France, Italy and Spain to endorse Hutton's ideas, and hence to reject catastrophism. The idea of uniformitarianism was also important in influencing the development of ideas in other disciplines. Work on the origin of the Earth's species by Charles Darwin and Alfred Wallace extended the idea into the biological sciences. The theory of evolution is based on the principle that the diversity seen in the Earth's species can be explained by the uniform modification of genetic traits over long periods of time. We have to remember that prior to the late 18th century many ideas about the development of the Earth's land surface were constrained by beliefs that the Earth was no older than 4004 BC and that features had been produced by divine catastrophic intervention. The 19th century was a period of gradual formulation of ideas for which Hutton and Lyell provided an extremely important foundation. In *The History of the Study of Landforms* (Chorley et al., 1964) Hutton and uniformitarian thinking appear in the first 30 pages and also on the last (page 641), thus demonstrating its pervading influence throughout the development of ideas in the 19th century. The next one hundred years saw the assimilation of uniformitarian ideas into geomorphological thinking up to the *Principles of Geomorphology* where Thornbury (1954) placed uniformitarianism as his first concept.

In the second half of the 20th century the significance of the concept was critically reviewed. Hooykaas (1959) contended that Lyell's uniformitarianism is not a single idea but embraces four related propositions, namely:

- *Uniformity of law* – the laws of nature are constant across time and space.

- *Uniformity of methodology* – the appropriate hypotheses for explaining the geological past are those with analogy today.

- *Uniformity of kind* – past and present causes are all of the same kind, have the same energy, and produce the same effects.

- *Uniformity of degree* – geological circumstances have remained the same over time.

With the title of his article 'Is uniformitarianism necessary?' Gould (1965) reduced these four propositions to two, suggesting that it was in effect a dual concept: first postulating uniformity of rates of geologic change, and second the time and space invariance of natural laws. He concluded that his first element, which is 'substantive uniformitarianism', should be abandoned because we know that rates of change have altered through geologic time and have been substantially altered more recently by human agency, and it also inhibits hypothesis formation. He argued that his second element, methodological uniformitarianism, belongs to science as a whole, is not unique to any single discipline and so effectively is now superfluous. Subsequently Shea (1982) compiled 'Twelve fallacies of uniformitarianism' because he contended that although uniformitarianism is widely recognized as *the* basic principle of geology, the literature contains many misleading statements as to what it really means. His 12 fallacies were that uniformitarianism:

1. is unique to geology;

2. was originated by Hutton;

3. was named by Lyell, who established its current meaning;

4. should be called 'actualism' because it refers to 'real' causes;

5. holds that only currently acting processes operated during geologic time;

6. holds that the rates of processes have been constant;

7. holds that only gradual processes have acted and that catastrophes have not occurred during Earth's past;

8. holds that conditions on Earth haven't changed much;

9. holds that Earth is very old;

10. is a testable theory;

11. is limited in both time and place;

12. holds that the laws governing nature have been constant through time.

Shea (1982; 1983) therefore concluded that 'Geologists should aban-
don the terms "uniformitarianism" and "actualism" because they are
fruitless, confusing, and inextricably associated with many fallacious
concepts. Instead, the fundamental philosophical approach of science should be recognized as basic to geology'. These arguments by
Gould (1965) and Shea (1982) have proved to be influential so that
the *Dictionary of Physical Geography* (Kennedy, 1985; 2000; 2006)
explained that in some formulations the term merely expresses a scientific research strategy in which the simplest explanation consistent
with the known evidence is adopted in order that it is a guiding tenet of
science and not a rule of nature. In this context Kennedy (2006: 141)
suggested that because it has nothing whatsoever to do with the 'uniformity' or 'gradualness' of the operation of processes, it is best summed
up by Sherlock Holmes's dictum: 'When you have eliminated the impossible, whatever is left, however improbable, must be the truth'.

3.2 Additional concepts

As the concept of uniformitarianism evolved other concepts emerged,
some directly associated as indicated above. Schumm (1985), noting
that Gould (1965) had suggested the term uniformitarianism should
be abandoned, commented that Europeans prefer the term actualism (Table 3.1). The definition of uniformity encountered difficulties
because our understanding is now very different from that in the early
18th century and also because it has been used by physicists and others
in the restricted sense of natural laws being permanent and universal,
so that under the same conditions a given cause will always produce
the same results. Therefore in relation to extrapolation Schumm (1985)
preferred several principles, including a principle of *accountability*
(continuity) as a statement of the conservation of matter and of the
related conservation of energy, and the principle of *optimality*, which is
the tendency of an open system to move towards optimum efficiency, a
condition of steady state or dynamic equilibrium. Thus Schumm (1985)
concluded that uniformity and actualism embrace four concepts:

1. Natural laws and physical constants that are permanent (philosophi-
    cal uniformity).

2. Matter (e.g. water, sediment) is not lost as it moves through a system
    (accountability).

3. Metaphysical explanations are not admissible and rates cannot
    exceed physical limits (simplicity).

4. Open systems tend to adjust toward a steady state (optimality).

Thus Baker (2012) concludes that Schumm's philosophy holds that extrapolations in geology largely derive from a concept of geological uniformity (actualism) and a specialized use of analogical reasoning.

Gradualism envisages slow, continuous processes and is an idea which is attributed to Hutton, maintained by Playfair, and elaborated by Lyell. This mode of thinking underpins Darwin's ideas on evolution, and his dictum 'nature does not make jumps' is a catchphrase for the school of evolutionary change (Huggett, 1999). Some writers (e.g., Kennedy, 2006) perceive gradualism as indistinguishable from uniformitarianism, seeing both gradualism and actualism emerging as versions of uniformitarianism.

The kind of **catastrophism** that prevailed before the early 19th century was displaced by some concept of uniformitarianism. However, in the mid 20th century there was a growing realisation that gradual evolutionary processes were, after all, punctuated by the effects of large catastrophic events. To distinguish what some have described as a revival of catastrophism, Dury (1980) imported *neocatastrophism* into physical geology and geomorphology from palaeontology. **Neocatastrophism** had been introduced by palaeontologists who were concerned with the sudden and massive extinctions of life forms in the marine fauna in the Permian, the dinosaurs in the Jurassic/Cretaceous, and the great mammals in the late Pleistocene. Dury cited examples, both general and specific, to exemplify a neocatastrophist viewpoint and cited the failure of the ice dam of glacial lake Missoula, which produced enormous features arising from the catastrophic discharge, as an excellent example of the need for a neocatastrophic explanation. As views of progressive and gradual shaping of the surface of the Earth had become well established it was difficult to assimilate catastrophic views, and Huggett (1988a) explored ways whereby the morphology of some landscapes should be recognized as being in large part fashioned by catastrophic and cataclysmic events (Huggett, 1988a, b), subsequently demonstrating the ways in which asteroids, comets and other dynamic events had been significant in Earth's history (Huggett, 1997b). Huggett (1988a) cites three main groups of hypotheses which involve some kind of terrestrial catastrophism: those involving a shift of the Earth about its axis of spin; those involving a shift of the axis of spin itself; and those involving bombardment by asteroids, comets and meteorites. Effects can be thought of as primary (the formation of tidal waves, shock waves, hurricanes, impact craters and changes of sea level), and secondary (which arise from primary effects and include climate change, extinction of organisms, volcanism episodes, and the cataclysmic change of landforms). The origin and development of catastrophism in the geoscience community have been tracked by using the bibliographic research tool Scopus to explore 'catastrophic' words in the Earth and planetary science literature 1950–2009 (Marriner et al., 2010). This analysis shows that the neocatastrophist mode became prominent in North America during the 1960s and 1970s before being more widespread in Europe after 1980, and was succeeded by an exponential rise in neocatastrophist research from the 1980s onwards. They attempt to explain this rise of neocatastrophism by highlighting seven non-exhaustive factors: (1) the rise of applied geoscience – encouraged by the impact of natural hazards and disasters; (2) inherited geological epistemology – the geological record of catastrophes closely mirrors the way in which stratigraphic boundaries have been significant in elucidating unconformities; (3) disciplinary interaction and the diffusion of ideas from the planetary to earth sciences – since the 1960s the impact craters on many solar system bodies have also been found on Earth; (4) the advent of radiometric dating techniques – the replacement of relative stratigraphies by absolute chronologies enabled the duration of events to be dated, showing that some were much briefer than would be expected under uniformitarian ideas; (5) the communications revolution – especially via the internet which has led to more international ideas and reduced national schools of thought; (6) 'webometry' and the quest for high-impact geoscience – large events make a 'good story' which few editors can resist; and (7) popular cultural frameworks – a heightened public perception of large-scale environmental disasters, also the importance of place (for example a New World perception contrasts with that in Europe). They conclude that there is no universally accepted definition of neocatastrophism, with episodism and convulsive events appearing as new axioms. Since the 1980s there has been an exponential increase in the reporting of disasters and also a significant downscaling of episodic geological events from the global to the regional and even to the local. Furthermore, they conclude that much of catastrophism's new-found popularity reflects not only the rigidities of the classic uniformitarian model (Baker, 1998) but also socio-political anxieties in the face of global change. Baker (1998) contrasts uniformitarianism, which uses regulative principles including simplicity, actualism and gradualism, with catastrophism, which is rooted in the view that Earth signifies its causative processes in landforms, structures and rock. Whereas uniformitarianism is tied to an early 19th century view of inductive inference, catastrophism hypotheses are generated retroductively. This remains relevant to modern science whereas outmoded notions of induction have been shown to be overly restrictive in scientific practice.

Other terms that have been used are included in Table 3.1. Step functional changes were referred to by Dury (1980) importing the idea of step function which has been used in science (and technology) since 1929 for a sudden discontinuous change or a mathematical function of a single real variable that remains constant within each of a series of adjacent intervals but changes in value from one interval to the next. This has been imported to account for the fact that changes over time are not always progressive and uniform but can sometimes be punctuated by

Table 3.1   Related concepts

Concept	Definition	Originator
Actualism	Explanation based on the understanding of present processes for postdiction and prediction	Schumm (1985) cites Jong (1966) and contends that it is important to replace the broad concepts of uniformitarianism and actualism with more specific principles of philosophical uniformity, accountability, simplicity and optimality
Gradualism	Profound change is the cumulative product of slow but continuous processes, sometimes used synonymously with uniformitarianism	The idea began with Hutton or Lyell but word not used in Lyell
Catastrophism	The morphology of some landscapes is in large part fashioned by catastrophic and cataclysmic events	William Whewell (1832)
Neocatastrophism	The effects of huge natural catastrophes supplementing the effects of gradual evolutionary processes shaping the Earth	Dury (1980)
Step functional change	A sudden discontinuous change	A term borrowed from science and technology, cited by Dury (1980)

sudden rapid changes (see Dury, 1980). The sudden jumps that character-ize switches from one equilibrium state to another are the steps in a step function. In a classic example this was implicit in the way in which Knox (1972) envisaged the phasing of environmental parameters in response to disruption and hence related to thresholds (see Figure 6.1).

3.3 Present significance

Uniformitarianism has been so significant in geomorphology and in the Earth and environmental sciences that it could be regarded as an overarch-ing paradigm. All disciplines have a set of defining practices, or paradigms, within which the science operates and, as the discipline develops, there are a series of paradigm shifts that arise during the evolution of thought in the discipline. Thus Gregory (2010: Table 2.5) listed paradigm shifts that have characterized study of the Earth's land surface and suggested that uniformitarianism/catastrophism qualify to be included. Until c.1800 a version of catastrophism was used to explain the Earth's land surface, but after a paradigm shift this was succeeded by uniformitarianism thinking until it was realized that explanations are rather more complex and that certain catastrophic events have contributed to the land surface so that a further paradigm shift needs to encompass neocatastrophism.

An oft-quoted illustration is the channelled scabland in Washington USA, site of one of the great controversies. Landforms on the Washington plateau, including what are now known to be giant current ripples up to 15m high, were thought by Bretz (1923) to be the result of catastrophic flooding after the drainage of Late Pleistocene proglacial lakes that filled the pre-existing valleys. However Bretz encountered great resistance to his ideas because some of the features produced were so large. It was much more recently, particularly as a result of work by Baker (1978a, b; 1981) aided by satellite imagery, that the catastrophic nature of landscape development was appreciated and it was realized that flood discharges could have been as large as 21.3 million $m^3.s^{-1}$. Elsewhere there are major features that are attributed to major catastrophic events, and examples are included in Table 3.2.

Table 3.2 Features attributed to major catastrophic events

Area	Major System	Comment
Cordilleran ice	Channelled Scabland	Flooding from Lake Missoula occurred between approximately 17,000 and 12,000BP (see Baker and Bunker, 1985)
Laurentian ice sheet	Great Lakes	The Great Lakes formed at the end of the last glaciation about 10,000 years ago, during deglaciation: glacial Lake Agassiz and the present distribution of the Great Lakes were created
Fenno-Scandinavian ice	Baltic Ice Lake	The Baltic Ice Lake c. 10,000BP was succeeded by the Yoldia Sea, Ancylus Lake and Litorina Sea as precursors of the Baltic Sea
	Urstromtaler Pradoliny	Blocked north flowing rivers including the Vistula so that major meltwater systems directed along pradolinas to the North Sea via the Elbe valley
Siberian ice	Southern Siberia, Altai Mountains	Ice-dammed lakes of south Siberia with total area at least 100.10^3km^2 dammed by pulsating glaciers and outburst discharges were comparable with those from Lake Missoula (see Rudoy, 1998)
Southern Britain	Formation of English Channel	Two catastrophic floods from ice dammed North Sea towards the Bay of Biscay created the Straights of Dover and English Channel c. 450,000 years ago (see Gibbard, 2007)

Uniformitarianism was an influential concept during the development of geomorphology. It perhaps now remains as the scientific research strategy in which the simplest explanation consistent with the known evidence is adopted and is, as mentioned at the beginning, reminiscent of Occam's razor.

FURTHER READING

Baker, V.R. (1998) Catastrophism and uniformitarianism: logical roots and current relevance in geology. In D.J. Blundell and A.C. Scott (eds), *Lyell: The Past is the Key to the Present.* London: Geological Society (London). pp. 171–82.

Dury, G.H. (1980) Neocatastrophism? A further look, *Progress in Physical Geography,* 4: 391–413.

Marriner, N., Morhange, C. and Skrimshire, S. (2010) Geoscience meets the four horsemen? Tracking the rise of neocatastrophism, *Global and Planetary Change,* 74: 43–8.

Record, J.A. (1997) Introduction to Charles Lyell *Principles of Geology.* London: Penguin. pp. ix–xliii.

Shea, J.H. (1982) Twelve fallacies of uniformitarianism, *Geology,* 10: 455–60.

TOPICS

> 1. How necessary is an understanding of uniformitarianism for contemporary students of geomorphology?

 ## WEBSITE

For this chapter the accompanying website **study.sagepub.com/ gregoryandlewin** includes useful articles in *Progress in Physical Geography*. References for this chapter are included in the reference list on the website.

4

LANDFORM

Landform has been basic to the study of geomorphology since the late 19th century, and form component definition evolved as a central concept until the second half of the 20th century when remote sensing, GIS, DEMs and geomorphometry allowed more rigorous quantitative procedures, although this also involved a review of first principles. Land form, as land shape, may have received more attention than landforms, involving their genesis, and awareness of natural kinds reminds us how both are dependent upon human perception.

The word geomorphology means to write about (Greek *logos*) the shape or form (*morphe*) of the Earth (*ge*), so that the simplest definition of the discipline is the scientific study of landforms. Every scientific discipline has a central focus and for geomorphology the landform is so central to the discipline that many geomorphology books do not define it! Landforms have been portrayed as natural features of the Earth's surface, as discrete geomorphological units defined by surface form and location in the landscape, or as part of continuous or multi-faceted terrain. Thus identified, units or elements may be categorized by characteristic physical attributes such as shape, elevation, slope, orientation, stratification, rock exposure, and soil type, and they can range from large-scale features such as plateaus to small-scale features such as fans. Each landform on the surface of the Earth occupies a particular scale in space and time, as with Ahnert's (1981) illustration developed in a different way in Figure 4.1 with a hierarchical classification shown in Table 4.1.

Although the Preface of Volume 1 of *The History of the Study of Landforms* (Chorley et al., 1964: xi) states that 'After about 1860 the study of landforms became part of both geology and physical geography and was later known as physiography or geomorphology', it is not easy to discover exactly when landform types first became basic for the science of geomorphology. Baron Ferdinand von Richthofen (1833–1905), who trained in geology and geography at Breslau (now Wroclaw), in 1886

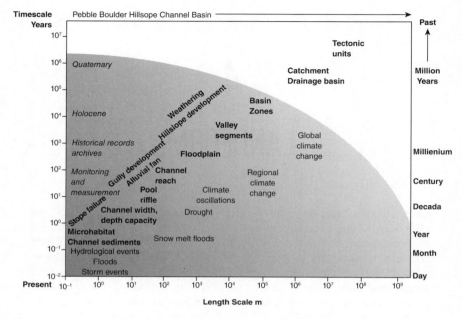

Figure 4.1   Landforms in relation to space and time

Table 4.1   Hierarchical classification of geomorphological features (time and space scales are approximate; developed from Chorley et al. (1984) and Baker (1986): http://geoinfo.amu.edu.pl/wpk/geos/geo_1/GEO_CHAPTER_1. HTML)

Typical Units		Spatial Scale km^2	Time Scale Years
Continents		10^7	10^8–10^9
Physiographic provinces, mountain ranges		10^6	10^8
Medium and small-scale units, domes, volcanoes, troughs		10^2–10^4	10^7–10^8
Erosional/ depositional units	Large scale, large valleys, deltas, beaches	10–10^2	10^6
	Medium scale, floodplains, alluvial fans, cirques, moraines	10^{-1}–10	10^5–10^6
	Small scale, offshore bars, sand dunes, terraces	10^{-2}	10^4–10^5
Geomorphic process units	Large scale, hillslopes, channel reaches, small drainage basins	10^{-4}	10^3
	Medium scale, slope facets, pools, riffles	10^{-6}	10^2
	Small scale, sand ripples, pebbles, sand grains, striations	10^{-8}	10^{-2}

published what may have been the first systematic textbook of modern geomorphology (Fairbridge, 1999). Landforms gradually became assimilated into the scientific literature (**Table 4.2**) on the land surface of the Earth (Gregory, 2010), notably in the course of exploration of the American West and through the contributions of William Morris Davis (1850–1934) that formalized the importance of the landform as a genetic entity (Davis, 1900: 158).

Landforms provide the building blocks of landscapes (4.1), whether as genetically defined entities, or as surface form units. This involves classification, additional concepts such as associations and hierarchies (4.2), and a variety of different perceptions (4.3).

4.1 Building blocks of landscape

How do we identify and describe the basic component of the Earth's surface, analogous to the way that pedologists recognize the soil profile and ecologists distinguish the habitat? Language includes words for particular Earth surface features. English nouns such as mountains, plains, valleys and plateaux have equivalents in other languages, but some languages include words for Earth surface characteristics of the environment distinctive to their particular country, so that a language of place (Mead, 1953) reflects how some vocabularies have unique words for particular features. In Russian there are words for types of valley, like *balki*, which cannot easily be translated into English; many descriptive words have become used for landforms in the way that corrie, cirque or the Welsh word *cwm* for armchair-shaped hollows have become accepted as features of glacially eroded landscapes, and tors are landforms found in periglacial and tropical regions. Many of the words used for landforms have become technically underpinned (**Table 4.3**); many others, such as valleys and hills, are common usages and are not precisely defined.

Words are not sufficient so that mapping, profiling and now three-dimensional visualizations have been employed to show the character and extent of landforms. Two approaches have been used: a morphological approach concerned with **land form,** and a genetic approach recognizing **landforming.** Although the Earth's land surface may be regarded as one continuously variable interface, it is usually recognized that this encompasses patches or overlapping palimpsests of elements that may be in sets or sequences or of different origin (glacial, fluvial, etc.). A common approach has also been to identify the smallest units that can be recognized, the undivided flat or slope, given that all land surfaces are composed of a jigsaw of such morphological units (Linton, 1951). Such units of relief that Linton (1951) characterized as the electrons and protons of which physical landscapes are built have much in common with the 'site', originally described as 'an area, which appears for all practical purposes to provide throughout its extent similar conditions as to climate, physiography, geology, soil' (Bourne, 1931). Landforms are composed of such morphological units, which may then combine to make larger-scale entities. When morphological units were first recognized, electrons and protons were thought to be the basic units of matter, but as with matter morphological units can now be subdivided into smaller constituents – the particle, or the pixel. The pixel, a term contraction from 'picture element', is the basic unit in a grid of pixels or raster as used early in television but now applied widely to imagery compilation. For landforms, we may see alluvial phenomena range upwards in scale from individual particles through landform units (such as levees) to alluvial complexes or ensembles sometimes characterized as alluvial 'architecture' (Lewin and Ashworth, 2013). Landforms therefore encompass a great range of spatial scales from the undivided continent to the minutest slope element on the Earth's surface (Figure 4.1, Table 4.1).

Morphological maps can be produced to show the distribution of slopes, defining the land surface in terms of basic flats and slopes (though at a scale larger than rock or soil particles). Such mapping schemes, effectively slope maps, are useful because areas of slope of particular angles can relate directly to land use practices, to the angles at which agricultural implements can operate, or to the slope angles at which mass movements occur. Mapping basic morphological components of the Earth's surface required many hours of field work (e.g., **Figure 4.2**) but two major developments have revolutionized the depiction of the form of the land: remote sensing and geographical information systems (GIS). Technological developments enabled dramatic progress and the availability of new data sources providing greater spatial resolution has allowed new insights and rapid mapping to be performed, organized after the 1960s within the framework of a Geographic Information Systems (GIS) defined as the collection, analysis, storage and display of data spatially referenced to the surface of the Earth. There is clearly a family relationship with pixelated imagery, but it also allows the identification of patterns and relationships between phenomena and processes (Oguchi and Wasklewicz, 2011). For example, numerical land classifications based on 1-km grid squares can combine many aspects of environmental character without being use-specific. Digital Elevation Models (DEMs), for larger or smaller scale resolution, can now be used to model the shape of the land surface and are integral parts of GIS (**Figure 4.3**). Rather than deciding scales or the identity of landform types (like moraines or point bars) *a priori*, spatial resolution may be set numerically by permitted pixel or sampling grid size. Experimentation to decide on the appropriate resolution for particular purposes is possible. A digital data framework for the organization of spatial data and the co-registration of data into a single geodetic reference system have both been very significant, and the avail-ability of DEMs has been at the forefront of much recent research (Smith and Pain, 2009). Whilst DEMs have been generated from contours and aerial photos for some time, the advent of routine space-based data col-lection through photogrammetric processing using dedicated fore/aft sen-sors (e.g., SPOT 5, ASTER) and interferometric synthetic aperture radar (InSAR; Rosen et al., 2000) has enabled great progress to be made. The move towards DEMs of higher spatial resolutions and vertical accura-cies, together with the advent of Light Detection and Ranging (LiDAR), has given useful results for a whole range of landform studies. Further developments can be achieved by interactive 3D visualization based upon multiple elevation surfaces with cutting planes used to analyse landscape structure based on multiple return (LiDAR) data. Multiple surfaces and 3D animations can introduce novel concepts for visual analysis of terrain models derived from time-series of LiDAR data using multi-year core and envelope surfaces (Mitasova et al., 2012).

In addition, since 1994 the advent of Global Positioning Systems (GPS) has enabled the determination of a specific location anywhere on the surface of the Earth, employing a navigation system with a constella-tion of 24 orbiting satellites, facilitating a revolution in the identification of landform global location.

Although the identification of genetic landform types requires an expe-rienced observer, there have been attempts to develop automated and semi-automated techniques for landform identification or feature extrac-tion evolving to a research area of **geomorphometry**, or quantitative land surface analysis (Oguchi and Wasklewicz, 2011). Geomorphometry is the science of quantitative land-surface analysis, with a dedicated inter-national society (**Box 4.1**). Progress in geomorphometry has included consideration of what landform actually is (Evans, 2012) and of how the land surface can be defined from an overabundance of data. Addressing operational definitions, a hierarchical taxonomy of fundamental geo-morphometric variables has been proposed (Evans and Minar, 2011) composed of field variables and object variables (**Table 4.4**). New ways of characterizing the land surface require developing novel methods for the classification and mapping of landform elements from a DEM based on the principle of pattern recognition rather than differential geometry. One approach is the concept of **geomorphon** (geomorphologic phono-types) (Jasiewicz and Stepinski, 2013), a relief-invariant, orientation-invariant, and size-flexible abstracted elementary unit of terrain. This is expressed in terms of local ternary pattern that encapsulates morphol-ogy of surface around the point of interest. Geomorphons enable terrain analysis without resorting to differential geometry, and a collection of 498 different geomorphons constitutes a comprehensive and exhaustive set of all possible morphological terrain types (Stepinski and Jasiewicz,

2011). This can give a general-purpose geomorphometric map by generalizing all geomorphons to a small number of the most common landform elements. Such maps are suggested to be a valuable new resource for both manual and automated geomorphometric analyses (Jasiewicz and Stepinski, 2013).

Geomorphometry and geoinformatics (**Box 4.2**) now permit the production of morphological maps very rapidly and accurately using consistent criteria. Criteria definitions are required, but sufficient recourse to the knowledge gained from earlier, field-based literature is also needed. There is some similarity here to approaches to plant classification in biology – initially morphological and based on appearances, but increasingly involving genetics as a means towards understanding the underlying structures to life forms. An interesting contrast with older 'geomorphological maps' that plot the distribution of genetic landform types, such as moraines or point bar ridges and swales, is that there are no blank areas left between such mapped units. Genetically speaking, glacial, fluvial and aeolian landscapes can often be 'feature-free' sloping terrains of sedimentary and rock surfaces even though they have been produced in such process domains, as well as having sets of defined forms such as barchan dunes or U-shaped valleys.

Morphological maps in themselves may not reflect the origin of the surface unless a unique form signature can be determined. The alternative genetic approach usually characterizes surface morphology, together with landform origin, dates for each section of the land surface, and indications of rock types, sediments and soils beneath the surface. These are important diagnostic tools in process studies. Not all requirements may be achieved in a single map, and academic papers commonly have maps and sections showing keyed elements to suit their own diagnostic purposes. General geomorphological maps produced in particular countries have also had their own emphases. In one of the most successful schemes in Poland, maps were produced at the scale of 1:50,000. Enthusiasm for general geomorphological maps has been limited, because their production, certainly for whole countries, has been prohibitively expensive with constant revision and updating required. Recently the advent of remote sensing sources has enabled a renaissance of geomorphological mapping (Smith and Pain, 2011) surveying remote regions, in greater (topographic) detail, over increasingly smaller time periods, accompanied by the emergence of aerial and terrestrial datasets enabling new applications. These may be very information-rich, but require interpretation skills and procedures to interpret them.

Taken altogether, with the availability of new tools such as satellite imagery, global positioning systems, digital elevation models and GIS, it has been possible to have a more effective approach to the acquisition, storage and display of geomorphological features. Geomorphologists can produce geomorphological models, consisting of land surface 'objects', organized into hierarchically arranged classes with spatially variable properties and geometric relationships (Dramis et al., 2011). Specific developments continue to be made. The geomorphons noted above were used for a 30x30 m cell geomorphometric map generated for Poland (Jasiewicz and Stepinski, 2013). Elementary forms or land elements can be grouped together into functional regions (landforms) such as 'hill sheds' (Evans, 2012). A so-called InterIMAGE interpretation strategy proved to be effective for the extraction of landforms (Camargo et al., 2012). By constructing a new legend at a scale of 1:10,000, combining symbols for hydrography, morphometry/morphography, lithology and structure with colour variations for process/genesis and geologic age, it has been possible to produce a 'geomorphological alphabet' (Gustavsson et al., 2006) that can be used to portray landscape configuration and illustrate the reconstruction of its temporal development.

Recognizing that geomorphological mapping plays an essential role in understanding Earth surface processes, geochronology, natural resources, natural hazards and landscape evolution, new spatio-temporal data and geo-computational approaches now allow Earth scientists to go far beyond traditional and subjective mapping, permitting a quantitative characterization of landscape morphology and the integration of varied landscape thematic information that extends beyond pure form (Bishop et al., 2012). Consequently recent progress in landform identification using new sources and techniques prompted a suggestion (Smith and Pain, 2011) that geomorphology really is an 'interface' discipline – not just physically, in the sense of studying the Earth's land-air or land-water surface, but also between pure and applied sciences that seek to derive greater benefit from integrating Earth's surface processes and landforms into their analyses. Geomorphology can be a necessary key integrating discipline for the geosciences, analogous to geological mapping as a key underpinning resource for societal development.

4.2 Classification, hierarchies and associated concepts

Since landforms were first systematically identified (Table 4.1), classifications have been necessary, augmented from geomorphometry, remote sensing and GIS, all enabling relationships of landforms to be analysed, and their association with other concepts understood. The six major ways of classifying landforms (Table 4.5, column 2) suggested by Beckinsale and Chorley (1991: Chapter 11) overlap somewhat and so four major categories are developed in Table 4.5 in order to indicate the present status of different approaches.

Table 4.5 Approaches to the classification of landforms

Type of Classification		Example, Citation	Current Status
1. Genetic	Encyclopedic	Peschel (1870) see Table 4.1. Was basis for subsequent recognition of range of landforms classified according to genesis. Recognized from mid 20th century on geomorphological maps.	Landforms associated with exogenetic geomorphic processes (weathering, slope, fluvial, coastal, Aeolian, glacial, periglacial), with endogenetic geomorphic processes (tectonic, volcanic), and with structural controls (karst): 498 different geomorphons constitute a comprehensive and exhaustive set of all possible morphological terrain types.
2. Morphological	Subdivision	Site: an area with similar local conditions of climate, physiography, geology, soil (Bourne, 1931). Nature offers two inescapable morphological units: at the one extreme the undividable flat or slope, at the other the undivided continent (Linton, 1951; Mabbutt, 1968).	Soil classification approaches terrain segmentation (Romstad and Etzelmüller, 2012)
	Accretion	Hierarchy of divisions recognized by Unstead, 1933: stows→tracts→regions. Wooldridge (1932): slopes and flats; the ultimate units of relief are flats and slopes (Linton, 1951; Gregory and Brown (1966): morphological units.	Geomorphic provinces Geomorphons Ecological patches (Bravard and Gilvear, 1996)
3. Process based	Drainage basin hierarchies Slope sequence Sedimentary architecture	Stream ordering: Horton (1945); Strahler (1952); and subsequent methods. Nine unit model (Dalrymple et al., 1969). Unit hierarchies (Miall, 1996).	Hill sheds (Evans, 2012) and landform geomorphometry. GIS. Applied to alluvial landforms (Lewin and Ashworth, 2013).
4. Applied	Practical	Land systems: an area with a recurring pattern of topography, soils and vegetation, CSIRO (Christian and Stewart, 1953).	Terrain units. Glacial, paraglacial land systems. Hazard and risk zoning maps
	Complex regionalization		Landslide distribution zoning maps (Calvello et al., 2013).

Initially labelled as encyclopaedic in the late 19th century, the recognition of different landforms according to origin developed so that a broad distinction could be made between those associated with exogenetic and with endogenetic processes as well as some in which bedrock geology is dominant, as in the case of karst landscapes. This potentially provides at least 10 broad categories, as reflected on many geomorphological maps. With developments in GIS and geomorphometry it has been possible to progress from human conceptualization with introduced subjectivity and bias with respect to the selection of criteria for terrain segmentation and placement of boundaries, to new spatio-temporal data and geo-computational approaches that now go far beyond traditional mapping. This permits a quantitative characterization of landscape morphology and the integration of varied landscape thematic information (Bishop et al., 2012) to produce geomorphological information about the land surface and landforms, but including additional information alongside it.

A second group of approaches essentially uses morphological data, categorizing surface form rather than origin, and this includes flats or slopes and has much in common with the recognition of site (Table 4.5) deriving from Bourne (1931). Recurrent patterns of spatial variation included the catena concept (Milne, 1935) expressing the way in which a topographic sequence of soils of the same age, and normally on the same parent material, can occur in landscape usually reflecting differences in relief/slope and drainage – an arrangement which others have described as a toposequence (Bates and Jackson, 1980). Following this approach from the viewpoint of the soil scientist, there are now many others involving landform by soil scientists and soil surveys, such as the Canada Soil Committee 1976 (www.pedosphere.ca/resources/CSSC3rd/chapter18.cfm) and the EU Joint Research Centre, European Soil Portal, 2012 (http://eusoils.jrc.ec.europa.eu/projects/landform/). Described as 'subdivision' by Beckinsale and Chorley (1991) this was complemented by 'accretion' whereby hierarchies of physical regions were identified, as exemplified by Fenneman (1931, 1938) in two substantial books identifying the landform regions under the heading of the physiography of the eastern and western United States. The contemporary manifestation of this is geomorphic provinces (e.g. Graf, 1987). Such approaches generally involve form hierarchies, with different levels that may be decided by the aggregation of lower-level units, or the subdivision of higher-level forms downwards ('lumping' or 'splitting').

A third group of approaches is based on processes (Table 4.5), first in the drainage basin, seen as the fundamental geomorphic unit (Chorley, 1969) – the area drained by a particular stream or drainage network and delimited by a watershed. It is a functional dynamic response unit from which outputs of water, sediment and solutes reflect the characteristics of the drainage basin that acts as the transfer function, and has been employed in recent GIS approaches. For slopes a nine-unit hypothetical landsurface model (Dalrymple et al., 1969) showed how nine particular slope components could occur on landsurface slopes anywhere in the world, with each component associated with a particular assemblage of processes. A similar approach was applied to pedogeomorphic research (Conacher and Dalrymple, 1977) where a simple five-unit slope may be sufficient (e.g., Birkeland, 1984). Alluvial systems may similarly be seen as both hierarchical and consisting of meso-scale elements such as channel bars, levees, overbank deposits and infilling palaeochannels, with all being developed simultaneously but at different rates (Lewin and Ashworth, 2013).

This 'process-based' approach follows a much earlier, and now more controversial, one adopted by W.M. Davis who characterized streams as 'consequent', 'subsequent', 'obsequent' and 'resequent' according to their sequence and origin in his theoretical cycle of erosion. A difficulty has been that such nomenclature depended on inference about landscape evolution rather than being readily determined from observable stream attributes. If an alternative evolutionary model becomes preferred, then the form elements require an identity revision, with some confusion between observation and interpretation.

A fourth category of applied approaches (Table 4.5) includes the **land system** developed by Christian and Stewart (1953) as areas with a recurring pattern not only of topography but also of soils and vegetation providing an approach for resource evaluation. Resource surveys in undeveloped parts of Australia and Papua New Guinea, initiated in 1946 by the Australian Commonwealth Scientific Industrial Research Organization (CSIRO), originated this approach. A further applied approach described as complex regionalization is illustrated by Russian work which includes recognition of the urochischa as a basic physical-geographical unit of landscape with uniform bedrock, hydrological conditions, microclimate, soil and meso-relief (Ye Grishankov, 1973) which could then be grouped into progressively larger units often characterized according to their use and potential and used for land evaluation. Whereas landscape ecology is the study of pattern and process at the landscape scale (Forman, 1995), focusing on what systems in the landscape can generally be used for, landscape evaluation is the estimation of the potential of land for specific kinds of use which can include productive uses such as arable farming, livestock production and forestry, together with other uses that provide services or benefits such as water catchment areas, recreation, tourism and wildlife conservation (Dent and Young, 1981). Such approaches have been refined with the advent of information systems (Cocks and Walker, 1987), advanced developments in remote sensing and the development of geographical information systems (e.g., Heywood et al., 1998). Two major contemporary developments have occurred: first in geomorphometry and GIS, and second through the recognition of specific geomorphological land systems (see Chapter 2, pp. 16–17) especially for glacial and paraglacial landscapes. Thus six paraglacial landsystems were identified (Ballantyne, 2002a): rock slopes, drift-mantled slopes, glacier forelands, and alluvial, lacustrine and coastal systems; each containing a wide range of paraglacial landforms and sediment facies.

Many potential links exist between the four major categories of Table 4.5, such as ecological patches forming a mosaic connected by corridors in any scale of landscape that can be employed in hydrology, analogous to a patchwork of geomorphological units nested at different scales (Bravard and Gilvear, 1996).

Landforms should be seen in the context of place and landscape. Place is used to refer to that particular part of space occupied by organisms or possessing physical environmental characteristics (Gregory, 2009), whereas landscape comprises the visible features of an area of land, including physical elements such as landforms, soils, plants and animals, weather conditions, and also any human components, such as the presence of agriculture or the built environment. Physical places, as enshrined in place names or types of landscape, are not easy to define but progress was made by recognizing physical or natural regions. Phillips (2001) contends that historical and spatial contingencies are responsible for the character of places. Historical contingency means that the state of a system or environment is partially dependent on one or more process states or upon events in the past, arising from inheritance, conditionality and instability: inheritance relates to features inherited from previous conditions (see Chapter 15). Conditionality is when development might occur by two or more different pathways according to the intensity of a particular phenomenon, for example whether a threshold is exceeded to instigate different trajectories of development. Instability refers to dynamical instabilities whereby small perturbations or variations in initial conditions vary or grow over time giving divergent evolution. Spatial contingency occurs where the state of an Earth surface system is dependent on local conditions that relate to local histories, landscape spatial patterns and scale contingency.

4.3 Contemporary perceptions of reality and interpretations

The identification of physical environments is now realized to be culturally determined: so do people from different cultures see physical landscape, and therefore landforms, in the same way? Thus Harrison et al. (2004: 10) contended that 'landforms have traditionally been seen as discrete entities (as things in themselves). Geomorphological maps employ solid black lines around landforms, yet in the field the boundaries between (and within) landforms are often very far from clear. The identification of geographical landforms, therefore, involves a clear set of assumptions not only about the nature of landform, but also about its history (both as a landform and as an intellectual category, since these are intertwined)'. Furthermore there are many cases when interpretations of landform have changed according to scientific thinking at the time, as in the case of landscape elements identified in terms of the Davisian cycle of erosion as well as later interpretations involving planation surfaces and residuals of variously identified origins (Table 4.6). At the other end of the spectrum is use of the term 'rock glacier'; Allison and Brunsden (2008) showed how 21 terms from 33 authors were utilized until just the one term became generally accepted.

It is quite generally accepted that scientific disciplines divide the particulars they study into *kinds*, which are groupings or orderings that do not depend on humans. Theorization about kinds may follow (http://plato.stanford.edu/entries/natural-kinds/#NatKinChe) as part of essentialism, a general theory of natural kinds in philosophy. Essentialism concerning natural kinds has three main tenets: first, all and only the members of a kind share a common essence; second, this essence is a property, or a set of properties, that all the members of a kind must have; and third, a kind's essence causes other properties associated with that kind. The essence of the natural kind 'gold', for example, is gold's atomic structure (Ereshefsky, 2009). Richards and Clifford (2011) suggest that the philosophical issue is whether these categories and the classificatory structures of which they are a part are 'real' (i.e., are 'natural kinds'; see Rhoads and Thorn, 1996), or simply convenient mental constructs to impose some degree of regularity on the apparently diverse character of surface forms, with implications for the manner of enquiry and type of explanatory process which follows the initial description (Harrison, 2001). Is the landscape *naturally* constructed of discrete entities for which we *require* names – drumlins, cirques, barchans, yardangs, inselbergs, etc. – or is it simply a continuous 3-D surface, to some of whose topographic attributes we *arbitrarily* assign these names? Furthermore, as shown in the next chapter, the notion of equifinality implies that a given landform may result from more than one process regime or process history.

Thinking about landforms as natural kinds underlines the need for the description and classification of landforms to be more detailed, rigorous, and genetically based. However, genetic interpretations are liable to change (Table 4.6), so that landform identity can change also. And what may have been identified in effect as 'kinds' may not collectively cover the whole Earth surface. Without digressing too far to

Table 4.6 Examples of changing interpretation of particular landforms

Landform	Original Interpretation	Developments and Current Interpretation
Erosion surface/ planation surface	Erosion surface used especially in Britain in mid 20th century to describe flattish plain produced by subaerial erosion. Often reconstructed from small remnants in the landscape.	Planation surface subsequently preferred term because such surfaces could be produced by range of processes, including marine erosion, and usually regarded as the product of an erosion cycle or a prolonged period of erosion under particular erosional conditions.
Glacial drainage channels	In the first half of the 20th century most channels interpreted as overflow channels and explained as produced by drainage overflowing from proglacial lakes.	Research on contemporary glaciers enlightened interpretation of former glacial drainage systems that were appreciated to be composed of channels that flowed on, in and under ice as well as at its margins. Hence the term 'glacial drainage channel' was employed to encompass a range of supraglacial, englacial, and subglacial routes.
Tors	In the first half of the 20th century regarded as weathering residuals typical of areas such as Dartmoor in the UK.	Subsequently the subject of debate because it was appreciated that they could be produced as integral parts of deep tropical weathering or during periglacial conditions.
Pediment	Originally applied by G.K.Gilbert (1880) to alluvial fans on the margins of Lake Bonneville, Utah.	Now thought of as smooth concave upward erosion surface that is part of the piedmont zone in arid and semi-arid areas. May have alluvial cover and have also been recognized in temperate areas.
Dry valleys	Until mid 20th century thought to be confined to limestone areas, and possessing all the characteristics of river valleys but no stream channel evident.	Later realized that dry valleys not confined to limestone outcrops, can occur on other lithologies, and reflect former more extensive drainage networks.

consider such questions as 'Do mountains exist?' (Smith and Mark, 2003), we have to remember the difference between land form and landforming approaches in the past and the need to understand how landform, materials and processes (considered in the next chapter) are integrated to comprise geomorphological understanding. And if we do not appreciate just what we mean by the 'form' of the land sufficiently, how can we suggest how landforms should be remodelled or designed as an integral part of landscape conservation (Gray, 2009)?

FURTHER READING

Evans, I.S. (2012) Geomorphometry and landform mapping: what is a land-form?, *Geomorphology,* 137: 94–106.

Jasiewicz, J. and Stepinski, T.F. (2013) Geomorphons – a pattern recognition approach to classification and mapping of landforms, *Geomorphology,* 182: 147–56.

Murray Gray, M. (2009) Landscape: the physical layer. In N.J. Clifford, S.L. Holloway, S.P. Rice and G. Valentine (eds), *Key Concepts in Geography.* London: Sage. pp. 265–85.

Smith, M.J., Paron, P. and Griffiths, J. (eds) (2012) *Geomorphological Mapping: Methods and Applications.* London: Elsevier.

TOPICS

> 1. Add other examples of changing interpretations of landforms as suggested in Table 4.6.

 WEBSITE

For this chapter the accompanying website **study.sagepub.com/ gregoryandlewin** includes Figures 4.2, 4.3; Tables 4.2, 4.3, 4.4; Boxes 4.1, 4.2; and useful articles in *Progress in Physical Geography.* References for this chapter are included in the reference list on the website.

5

FORM, PROCESS AND MATERIALS

Approaches to a central concept of form, process and materials have focused on processes, landform evolution, and climatic geomorphology. Although these developed separately until the late 20th century a more holistic approach has recently brought them together, especially fostered by multidisciplinary research. It is now appreciated that advances in macroscale geomorphology have enabled large-scale landform developments to complement small-scale process research. Using covering law models of explanation, it is possible to recognize geographical, geophysical macro geomorphology, and historical approaches

Landform is the subject matter for geomorphology as the landform science, so that it follows that a central concept is the relationship of landform, process and materials. Although manifested in various ways over the last one hundred and fifty years years it has not frequently been stated explicitly. However, it has been presumed, although some have recognized a growing emphasis on the 'mutual interaction between form and process in the understanding of geomorphological systems' (Roy and Lane, 2003). This chapter provides a summary of interrelationships (5.1); indicates how distinct concepts emerged during the progressive development of geomorphology (5.2); and surveys the present position (5.3).

5.1 Relating form, process and materials

The relationship between these three characteristics of the Earth's surface can be summarized by a simple geomorphological equation, adapted from a physical geography equation (Gregory, 1978a) subsequently accepted by a number of writers including Yatsu (1992) and Richards and Clifford (2011). The equation was devised to indicate the way in which processes (P) operating on materials (M) over time t produce results expressed as landforms (F). In equation form it can be expressed as:

$$F = f(P, M) \, dt$$

Geomorphological investigations can be visualized as concerned with five levels of enquiry as summarized in Box 5.1.

BOX 5.1

A way of summarizing geomorphology (see Gregory 1985; 2000; 2009)

A geomorphological equation, indicating how *processes* operating on *materials* over time *t* produces *results* expressed as a landform, can be expressed as:

$$F = f (P, M) dt$$

Investigations can be made at five levels:

- *Level 1: study of the elements or components of the equation* – study of the components in their own right. Some studies can be focused on the description, which may be quantitative, of landforms, of soil or rock character, or of plant communities.

- *Level 2: balancing the equation* – study of the way in which the equation is balanced at different scales. At the continental level may involve the energy balance relating available energy for environmental processes to radiation, and moisture received in relation to locally available materials. Studies of this kind focus upon contemporary environments and upon interaction between processes, materials and the resulting landforms or environmental conditions.

- *Level 3: differentiating the equation* – includes studies analysing how relationships change over time. This requires reconciliation of data obtained from different time scales together with a conceptual approach. Includes impact of climate change and human activity which may be the regulator that has created a control system.

- *Level 4: applying the equation* – when research results are applied to problems, very often extrapolating past trends, encountering the difficulty of extrapolating from particular spatial or temporal scales to other scales for which information is required to address management problems.

- *Level 5: appreciating the equation* – involves acknowledging that human reaction to physical environment and physical landscape can vary between cultures, affecting how the earth's surface is managed and designed

5.2 A theme prevailing in the development of geomorphology and landform science

During the course of the development of uniformitarianism there was an alternation of catastrophism and gradualism, but once the focus on landform relationships with processes and materials was embraced a sequence of geomorphological approaches evolved. It was perhaps inevitable that such approaches would each have a major driver, such as erosion, landscape change or climate. Portrayed by Jennings (1973) as a geomorphological 'bandwagon parade', this can be thought of as a sequence of paradigm shifts (Gregory, 2010: 35, Table 2.5). The emphases on form, process and materials have changed over the years (Richards and Clifford, 2011), but understanding the developing sequence is necessary to appreciate how they each affect contemporary thinking in geomorphology. Whereas we are now familiar with extremely rapid communication in geomorphology (e.g., Gregory et al., 2013) different approaches originated when it took much longer to surmount the obstacles of different languages, there were comparatively few national and international scientific meetings, few scientific journals for the publication of research and dissemination of ideas, and few researchers in the community of scholars concerned with investigations of the surface of the Earth.

It is not easy for a student in the 21st century to comprehend the multifarious strands that have produced the geomorphology of today, so that we require a convenient way to encapsulate the inter-relationship of the major founding developments, and to set them into the context of developments in other disciplines which were so influential at the time, permeating geomorphology and other sciences. Table 5.1 shows the main strands of thinking identified to focus on form, process and materials and underpinning thinking in geomorphology. What is much more difficult is divining the links that have occurred between the several strands, and ascertaining the extent to which similar ideas developed independently in different places. However, a map of past activity of this kind is needed to ensure that the modern outburst of literature and communication is not oblivious of past contributions – we must not re-invent the wheel! There is practical value in knowing the historical development of geomorphology (Sack, 2002).

The inclusions in Table 5.1 provide a framework that can be augmented and extended. The three major themes – process, evolution and climatic variety – are integrated with external trends which include systems, developments in other disciplines such as hydrology, analysis of ocean cores stimulating research in Quaternary science, remote sensing and the availability of other techniques including GIS and cosmogenic dating. Each of the major themes is explained in detail elsewhere (e.g., Summerfield, 1991; Gregory, 2000, 2010) and expanded in later chapters (11, 15), so that a brief outline is provided here.

The process theme dates from when there was a debate about how the Earth's surface was fashioned, and when actualism and gradualism succeeded catastrophism (see Chapter 3). However the main strand derived from the work of G.K. Gilbert (1843–1918), one of the explorers of the American West, who from the 1880s were demonstrating the power of subaerial erosion in producing landforms. In his later work Gilbert used analogy with physical mechanics and studied landforms as manifestations of geomorphic processes acting on Earth materials (Sack, 1991: 30), with the result that he is now acknowledged to be a brilliant geomorphologist who published a remarkable investigation on the *Transportation of debris by running water* in 1914, anticipating many developments that did not occur until nearly fifty years later. Although there were some subsequent fluvial contributions it was not until 1964 with the publication of *Fluvial Processes in Geomorphology* (Leopold et al., 1964) that a new era of process investigation became widespread, emphasizing physical principles, dealing 'primarily with landform development under processes associated with running water . . . better future understanding of the relation of process and form will . . . contribute to, not detract from, historical geomorphology'. Parallel with the interest in fluvial processes were other strands: coastal, glacial, and aeolian, the latter stimulated by a book on *Physics of Blown Sand and Desert Dunes* (Bagnold, 1941).

The second theme, labelled evolution, and possibly influenced by Darwinian evolution (1859), was introduced in 1895 by W.M. Davis (1850–1934). With the benefit of hindsight we now realize that his approach gave insufficient attention to the formative processes operating, was essentially qualitative in approach, focused on parts of the land surface and ignored others, and did not have a sound scientific foundation (see Chapter 11, **Box 11.1**, **Table 11.1**); however, his work was very intelligible and persuasively presented. The essence of his approach, which appealed to persons with little training in basic physical sciences, was that landforms are a function of structure, process and time, and evolve through stages of youth, maturity and old age. This conceptual model was devised for a 'normal' cycle of erosion applied to temperate landscapes, but alternatives of arid and marine cycles were also proposed, and in the course of landscape evolution there could be accidents, either glacial or volcanic. Land surface was interpreted in terms of the stage reached in the cycle of erosion and came to be dominated by a historical interpretation concentrating upon the way in which landscapes had been shaped during progression through stages in a particular cycle, towards peneplanation. Followers of this approach therefore attempted to identify the stages of long-term evolution of landscapes – an approach later

termed denudation chronology (see Gregory, 2000: 38–42). A collection
of Davis's influential essays and papers (Johnson, 1953) included 12 educational essays and 14 physiographic essays. This shows the interest that
Davis had in geographical teaching, fulfilling the substantial need that
existed at the time, and the popularity of his approach was attributed to
12 reasons (Higgins, 1975) with simplicity the first! This approach was
followed by others (see Chapter 11), with that of Lester King formalized as 50 canons of landscape evolution (King, 1953: 747–50). These
approaches to landscape evolution largely concentrated on the ways in
which landscape had been fashioned in the later stages of geological time
(see Chapter 13), the Cainozoic, including the Palaeogene, the Neogene
and the Quaternary. Although the analysis of Quaternary glacial impacts
was being investigated during the 20th century, generating researchers in
glacial and later in periglacial geomorphology, it was with improvements
in dating that Quaternary Science really flourished and began to evolve
as a separate field.

A climatic focus probably had its origins in Russia where soil scientists such as Dokuchaev (1846–1903) and his student Sibirtsev identified broad zonal patterns of soils related to climate. Dividing the land
surface of the Earth into major zones as a basis for considering how
different landforms occur in world landscapes, climatic geomorphology found favour in Europe and Russia because it could embrace the
way that soil and vegetation types are associated with particular zones,
reflecting also the morphoclimatic zones recognized in France (Tricart,
1957). In qualitative terms, phenomena could be regarded as zonal
if they were associated with the climatic characteristics of latitudi-
nal belts, whereas azonal phenomena are non-climatic such as those
resulting from endogenetic processes; extrazonal phenomena are those
occurring beyond their normal climatic limits such as sand dunes on
coasts rather than deserts; and polyzonal phenomena are those which
can operate in all areas of the Earth's surface according to the same
physical laws. Such zonality provided the basis for 13 morphoclimatic
zones (Tricart and Cailleux, 1965; 1972). An energy balance foundation was used to provide a quantitative climatic basis for geographic
zonality. Subsequently, three generations of geomorphological study
were recognized (Büdel, 1963) in a system (see Chapter 11) introduced in Germany which became more widely known after a paper by
Holzner and Weaver (1965), and was also gradually refined to culmi-
nate in eight climato-morphogenetic zones. A scheme of nine morphogenetic systems was introduced in the USA (Peltier, 1950; 1975), each
distinguished by a characteristic assemblage of geomorphic processes,
stimulating interest in periglacial environments in particular.

It might appear that materials have attracted less attention than process and form, although an early geological approach to geomorphology

believed that many features, including remnants of erosion surfaces, could be ascribed to the control by lithology on surface form. Awareness of the importance of lithology was particularly evident in limestone areas and karst geomorphology was named from the Dinaric karst region of Slovenia where features had been recognized as early as 1893 (Cvijic′, 1893; Benac et al., 2013). In karst areas field monitoring of solution processes produced some of the earliest measurements of erosion in the mid 20th century and it was from these that many subsequent process investigations stemmed. Karst is best developed on carbonate rocks which occur on some 14% of ice free continental areas (Ford and Williams, 2011) but related features also occur on other soluble outcrops such as gypsum (e.g., Doğan and Özel, 2005) and rock salt. Extensive research since the classification by Cvijic′ (1893) has meant that processes and landforms (**Figure 5.1**) have been well documented. Ford and Williams (2011) argue that karst punches above its weight in geomorphology because caves contain deposits that can now be radiometrically dated for many millions of years, the dated geomorphic history for karst regions provides a time scale for the evolution of surrounding areas, and the palaeoclimatic record of speleothems is more accurate and precisely dated and of higher resolution than the records from deep sea or ice cores.

Material properties analysed in relation to process and form benefit from the range of techniques now available for the analysis and description of the characteristics of rocks and superficial deposits (**Table 5.2**). The high degree of variability of material properties inhibits easy incorporation into landscape models (see **Table 11.3**) but has also stimulated greater links between weathering research with soil science. Soil geomorphology has been identified as the integration of pedology and morphology (Gerrard, 1993), demonstrating general relationships between soils, weathering and geomorphology (Birkeland,1974), and a more recent emphasis upon theory and process of soil genesis in relation to geomorphology (e.g., Schaetzl and Anderson, 2005). The significance of variations in material properties can have considerable import as shown by the difference between warm and cold glacier ice, and different conceptual active layer systems can respond differently to climatic change or disturbance based on the thermal properties of the material and ice/water content (Bonnaventure and Lamoureux, 2013). It is now possible to consider material properties at very large scales related to tectonics (Koons et al., 2012) because the heterogeneity and anisotropy of material strength are fundamental aspects of active orogens so that the description of the strength field in terms of mechanical evolution can extend present Earth surface models, expressed in landscape geomorphometrics of anisotropy and spatial patterns of complexity. Thus the lack of detailed investigations of material properties is now being redressed with the techniques which have become available.

Each of the themes, concentrating on aspects of form, process and materials, provided concepts in the sense of 'abstract ideas, general notions or units of knowledge vital to the development of scientific knowledge', as defined in Chapter 1 (pp. 2–3). Each evolved influenced by external developments, which included the theory of evolution (Darwin, 1859); the growth of hydrology and especially the influence of Horton (1945); developments in Quaternary science including the foundation of INQUA in 1928 and revolutions in dating, especially deep sea cores and cosmogenic dating; and developments in ecology and in philosophy – and not least the impact of systems thinking.

5.3 A view of the present position

Although there are obviously many links between the approaches (5.2) to form, process and material relationships, the differences that emerged came to be regarded as timeless and time bound approaches (see Chapter 1) associated with the two quite different viewpoints of geomorphology identified by Strahler (1952) as dynamic (analytical) and historical (regional) geomorphology. Any discipline develops towards more and more specialist branches and at least 24 have been identified as recently employed in books and research papers. These branches could be classified (Gregory and Goudie, 2011a: Table 1.5) according to purpose (Quantitative, Applied, Engineering), analysis (process, climatic, historical, structural/tectonic, karst, anthropogeomorphology) and process domains (Aeolian, Coastal, Fluvial, Glacial, Periglacial, Hillslope, Tropical, Urban, Weathering, Soil-geomorphology or Pedogeomorphology, Mountain geomorphology, Extraterrestrial geomorphology, Seafloor engineering geomorphology), as well as multidisciplinary hybrids (Hydrogeomorphology, Biogeomorphology). In addition to the identification of such sub branches there have been other diversifying trends. For example, geomorphology has been described as becoming a more rigorous geophysical science, but also as becoming more concerned with human social and economic values, environmental change, conservation ethics, the human impact on environment, and social justice and equity issues (Church, 2010).

With the development of so many branches of geomorphology, and new ones continuing to be created such as ice sheet geomorphology (e.g., Fleisher et al., 2006), it is perhaps inevitable that a more holistic approach has been sought. Whereas the first part of the 20th century saw the emergence of several branches of geomorphology, the second part witnessed the creation of many more subdivisions – fragmentation that has been characterized as investigating more and more about less and less, the so-called fissiparist or reductionist trend. The 21st century is seeing the culmination of efforts to realize a more **holistic approach**, namely a return to the 'big picture', and a holistic approach is also advocated within many branches including coastal, glacial, arid and fluvial. A holistic approach essentially means relating to, or being concerned with, complete systems rather than with the analysis of component parts. The holistic trend has affected other disciplines (in geomorphology being compatible with a systems approach and able to reinforce it). It had been suggested (Baker and Twidale, 1991) that while commendable in spirit, progressive initiatives to establish research traditions in landscape evolution, climatic geomorphology and process studies all encountered fundamental limitations as unifying themes. Therefore they stressed the need for a 're-enchantment' of geomorphology, which could arise from a new connectedness to nature. With hindsight the differences between approaches have not appeared to be so great, and for example the approaches of Gilbert and Davis to geomorphology can be regarded not as mutually exclusive but instead as complementary (Small and Doyle, 2012). The new techniques available, providing opportunities for all branches, appeared when there was increasing awareness of the need to counter the greater specialist emphasis upon components of the land surface without sufficiently acknowledging the links between them. For example, linkages between components (e.g., Brierley et al., 2006) can emphasize the ways in which nested hierarchical relationships between compartments in a catchment demonstrate both connectivity and disconnectivity in relation to geomorphic applications to environmental management. As there has been a greater general awareness of environment, and hence with applications of geomorphology, so the holistic nature of many problems demanding solutions has been appreciated. Such requirements have encouraged multidisciplinary research so that, as with other environmental and earth sciences such as the interface of geomorphology and ecosystems ecology (e.g., Renschler et al., 2007), hybrid disciplines have been fostered, including ecogeomorphology and hydrogeomorphology. Multidisciplinary investigations have been encouraged and 'biogeosciences' are rapidly expanding (Martin and Johnson, 2012), with investigations over a wide range of temporal and spatial scales. Added to these trends has been the greater attention given to macroscale geomorphology triggered by significant advances in plate tectonics (Summerfield, 2000) which, coupled with advances in cosmogenic dating, has led to a renaissance in understanding the development of large-scale Earth landforms.

Such progress towards the replacement or at least supplementation of the reductionist methodologies so successful for the progress of physics in the last century by a new holism is suggested (Baker, 2011) to afford a prospect for transcending the long-standing divide between historical and process studies. But what are the ways available to achieve this?

Adopting a more holistic approach directs attention to more comprehensive explanation in geomorphology. One general way is to employ the Covering Law model of explanation, also known as the deductive-nomological model (DN model), a formal view of scientific explanation, used since the mid 20th century, and particularly associated with the philosophers Carl Hempel and Karl Popper. This deductive explanation follows from the operation of general scientific laws with initial condition statements or premises forming the *explanans* which elucidates what is described as the *explanandum*. The DN model proceeds from laws to statements of particular initial conditions to explanation (see Haines-Young and Petch, 1986; Rhoads and Thorn, 1996). Recognizing that the defining features of a scientific explanation rest on the operation of general scientific laws, awareness of deductive reasoning and of the existence of alternative approaches may be why geomorphology has been much more methodologically concerned and explicit since the mid 20th century (Small and Doyle, 2012). Using an approach of this kind, Richards and Clifford (2011: 55) provided a summary diagram demonstrating the ultimate unity of geomorphology (**Figure 5.2**) which requires general 'laws' concerned with the functional nature of landforms, the immanent properties of Earth surface processes, and the adjustment of form to process.

More specific ways to achieve a holistic approach have underlined reviews and statements about future needs and foci such as the US National Research Council's committee on Challenges and Opportunities in Earth Surface Processes (see Murray et al., 2009). Prompted by governments requiring society benefits from their research funding, there has been the perception that few environmental challenges are likely to be solved by a single discipline because integrated approaches and interdisciplinary collaboration are often required. Individual suggestions have also been made, such as that of Lang (2011), to develop a computer modelling framework – an 'Earth surface simulator' – which would combine process understanding and evolutionary information to provide a unifying platform comparable to GCM technology. This could represent dynamic process interactions, including interfaces to the lithosphere, biosphere and atmosphere.

It has been suggested that three alternative foci can now be perceived (Gregory, 2010): geographical, interpreting morphology and processes; geophysical macro geomorphology, concentrating upon the broad structural outlines (see Church, 2005; Summerfield, 2005b); and chronological historical/Quaternary, focused on the history of change. However, these are much more connected, perhaps requiring a holistic approach, than approaches established in geomorphology since the late 19th century. Specialisms have greatly increased, with scientific compartmentalization, so that for example understanding the advantages and limitations of particular dating techniques is in a different realm from understanding GIS procedures. But the understanding of landforms, for practical management purposes in particular, may require forms of overarching collaboration as much as a highly focused specialism.

FURTHER READING

The third volume in the series on the history of the study of landforms introduces climatic geomorphology and global changes; the two preceding volumes provide detailed background to other developments:

Beckinsale, R.P. and Chorley, R.J. (1991) *The History of the Study of Landforms or The Development of Geomorphology Vol. 3: Historical and Regional Geomorphology 1890–1950*. London: Routledge.

Brierley, G.J., Fryirs, K. and Jain, V. (2006) Landscape connectivity: the geographic basis of geomorphic applications, *Area,* 38: 165–74.

Ford, D.C. and Williams, P.W. (2011) Geomorphology underground: the study of karst and karst processes. In K.J. Gregory and A.S. Goudie (eds), *The SAGE Handbook of Geomorphology*. London: Sage. pp. 469–86.

Gregory, K.J. (2010) *The Earth's Land Surface*. London: Sage. (See especially Chapters 3, 4.)

Huggett, R.J. (2011) Process and form. In K.J. Gregory and A.S. Goudie (eds), *The SAGE Handbook of Geomorphology*. London: Sage. pp. 174–91.

Rhoads, B.L. and Thorn, C.E. (1996) Towards a philosophy of geomorphology. In B.L. Rhoads and C.E. Thorn (eds), *The Scientific Nature of Geomorphology*. Chichester: Wiley. pp. 115–43.

Richards, K. and Clifford, N.J. (2011) The nature of explanation in geomorphology. In K.J Gregory and A.S. Goudie (eds), *The SAGE Handbook of Geomorphology*. London: Sage. pp. 36–58.

Summerfield, M.A. (2005) The changing landscape of geomorphology, *Earth Surface Processes and Landforms,* 30: 779–81.

TOPICS

1. Search for other subdivisions of approaches in geomorphology: do they accord with the suggestions in Table 5.1?

WEBSITE

For this chapter the accompanying website **study.sagepub.com/ gregoryandlewin** includes Figures 5.1, 5.2; Tables 5.1, 5.2; and useful articles in *Progress in Physical Geography*. References for this chapter are included in the reference list on the website.

6

EQUILIBRIUM

The concept of equilibrium emerged in geomorphology once ideas of cata-strophism had been succeeded by the understanding that gradual land-forming processes were responsible for the shape of the Earth's surface, and the idea of a 'balance of nature' prevailed. This was expressed first through the graded profile of rivers for time-bound studies, and subse-quently by the definition of other types of equilibrium such as dynamic, quasi-, metastable and steady state equilibrium, to account for results obtained from observation. The contemporary interpretation is that equi-librium is a significant and useful concept but that it is not universally applicable. There is not necessarily a single or final equilibrium state, and equilibria can be visualized in different ways – including as a metaphor.

What we now think of as **equilibrium,** the state of balance created by a variety of forces so that the state remains unchanged over time unless the controlling forces change, had begun to emerge in the early 19th century, especially in studies of rivers. Equilibrium is associated with force in dynamics, with energy in thermodynamics, and with pure numerical behaviour in mathematics. In General System Theory, equilib-rium is derived from thermodynamics but applied, by analogy, almost exclusively to mass (Thorn and Welford, 1994). As a conceptual frame-work for geomorphology, equilibrium thinking emphasizes the relation between present form and process (Mayer, 1992).

The sense of a balance being achieved, as an equilbrium between force and resistance, may have originated in 1697 with Domenico Guglielmini (1655–1710), who has been characterized as the father of the science of river mechanics. Kesseli (1941) traced subsequent developments to include Surell in 1841 and Dausse in 1857, up to Gilbert in 1877. Such early work was gradually recognizing that the long profile of a river is concave upwards and that it proceeds towards a graded profile. Dausse (1872) extended the idea of equilibrium to slopes. Work by Comte Louis Gabriel Du Buat (1734–1809), a Lieutenant Colonel in the Royal Corps of Engineers, set out to discover the factors controlling and influencing the flow and windings of a river, and he not only described the *vitesse de regime* (stability, equilibrium or grade) he also had a view of river stages anticipating the Davisian cycle by a century. Studies of river torrents by the French civil engineer Alexandre Surell (1813–1887) suggested that streams erode their upper courses and deposit in their lower courses so that the river is working towards an adjusted slope which he described as a period of grade (Chorley et al.,1964) – a subject later identified as an aspect of equilibrium. However, it was in the late 19th century that it became more clearly evident. G.K. Gilbert produced the first treatment (Gilbert, 1877) by a geologist of the mechanics of fluvial processes, employing grade to express what others had thought of as equilibrium. Gilbert laid the foundations for understanding the nature of fluvial processes, including the comminution of material during transport and the balance between erosion and deposition: when the river neither eroded nor deposited he termed grade (see Chapter 11) as an ideal stage or final stage towards which all rivers were working. His emphasis on equilibrium also led him to consider (Tinkler, 1985: 143) the forces responsible for upwarping the shorelines of Lake Bonneville (Gilbert, 1890). In his later years he analysed data from flume studies which produced important contributions on the transportation of debris by running water (Gilbert, 1914). Gilbert always showed a preference for the study of modern processes rather than Earth history (Baker and Pyne, 1978) so that his contributions were greatly valued when processes began to be studied in detail after the 1960s.

In 1884 Le Chatelier (1850–1936) introduced what has now been referred to as Le Chatelier's Principle in Chemistry which, in essence, was that 'If to a system in equilibrium a constraint be applied then the system will readjust itself to minimise the effects of the constraint'. This principle was used in thermodynamics, and it initiated a new era in chemical education; the idea is now basic to self-regulation, negative feedback and homeostasis. Coincidentally, also in 1884, W.M. Davis (1884) first introduced the elements of his idea for the cycle of erosion, including equilibrium as diagnostic of progression in his cycle of erosion. Davis contended that no part of a river reached equilibrium during youth, but the approach of equilibrium between a river's transporting power and load was termed maturity, and when a river reached equilibrium throughout all its tributaries the land was reduced to a general plain. An important book, *Les Formes du Terrain*, produced by Lt. Col G. De La Noe of the Geographic Service of the French Army and Emm. de Margerie, included ideas on slopes reflecting concepts more like those of Gilbert than those of Davis (Chorley et al., 1964), serving to apply the notion of equilibrium to slopes as well as to rivers (De La Nöe and Margerie, 1888).

Such 19th century ideas were fundamental in affecting subsequent progress so that in the 20th century equilibrium concepts played a major role in geomorphology, although Kennedy (1994), in an article titled 'Requiem for a dead concept', questioned whether the concept of equilibrium is truly central to geomorphology. Either way it is important to appreciate how equilibrium progressed from early origins (6.1), how related concepts were developed and applied (6.2), and how this allows an understanding of the present status (6.3).

6.1 Developing equilibrium

In the early 20th century geomorphology was offered both the Davisian cycle and a Gilbert approach founded on investigation of processes; they embraced equilibrium ideas of rather different kinds (Chapter 5, pp. 48–49). The Davisian approach prevailed, dominated geomorphology until the 1960s, and fostered an interpretation of landscape in terms of structure, process, and stage or time. A simple cycle approach was offered which remained current in geomorphology for at least sixty years. The way in which this constrained the development of geomorphology within an historical approach has been well documented in the second volume of the history of geomorphology (Chorley et al., 1973) including reference to the way in which Davis was able to incorporate pre-existing concepts into his schema. In particular, he embraced the concept of grade (see Dury, 1966) and acknowledged the idea from Gilbert (**Table 6.1**) and the applicability to a condition of balance pertinent to both a stream and longitudinal profile (Davis, 1894). Grade was the subject for significant reviews so that Kesseli (1941) concluded that there is no time when graded conditions are attainable; he defined grade in terms of profile such that a graded stream should be taken as one without waterfalls or rapids. In contrast Mackin (1948) extended the theory of grade and provided a redefinition (**Table 6.2**) to which he added a diagnostic characteristic of a stream in equilibrium: 'any change in any of the controlling factors will cause a displacement of the equilibrium in a direction that will tend to absorb the effects of the change' – this is reminiscent of Le Chatelier's Principle (see above). By contrast, Dury (1966) concluded that the concept of grade is 'unserviceable both in the study of actual terrains and the theoretical analysis of landform generally' (see Chapter 11, **Table 11.3**).

A greater emphasis upon hydraulics and channel processes was encouraged by engineers: particularly after the design of irrigation canals in India (Lacey, 1930), the concept of channels being 'in regime' was developed. This was a state in which, despite a throughput of water and sediment, canals could be essentially stable and without serious problems arising from erosion or deposition – an engineering concept essentially of equilibrium. Lane (1955), also from an engineering point of view (**Table 6.2**), proposed a general expression useful for analysing problems of channel morphology, and subsequently employed it as the basis for channel adjustments in the form:

$$Q_s d \; \alpha \; Q_w S$$

where Q_s is the quantity of sediment, d is the particle diameter or size of the sediment, Q_w is the water discharge, and S is the slope of the stream. Later work developed the idea that the dimensions of natural channels were in effect adjusted to a 'dominant' discharge at approximately bank-full stage, another manifestation of equilibrium-based thinking.

With the advent of more process-based investigations including dynamic geomorphology (Strahler, 1952) (**Table 6.2**) it was suggested that 'despite difficulties of definition the concept of equilibrium is a useful one', and furthermore that it 'implies both an adjustability of the channel to changes in independent variables such as load and discharge and a stability in form and profile' (Leopold et al., 1964: 267). Developments of grade (De La Nöe and De Margerie, 1888) were also applied to hillslopes (**Table 6.1**) by introducing an equilibrium theory of erosional slopes (Strahler, 1950) and by identifying as many as 14 concepts of equilibrium, grade and uniformity (Young, 1970). These included the graded slope as one 'possessing a continuous regolith cover, without rock outcrops', and the profile of equilibrium as a 'slope on which the position of each point of the profile depends at any moment on that of all of the others'. The application of equilibrium nomenclature to glaciers has been specific: the 'equilibrium line' is the notional boundary, so that above the line accumulation exceeds ablation each year and below it ablation exceeds accumulation. However, equilibrium-line altitudes (ELAs) appear to be highly variable from year to year, there are few glaciers whose ELA has been directly measured (Braithwaite and Muller, 1980), and the relationship between glacier ELAs and climatic variables such as precipitation and air temperature is complicated in high-mountain environments (Benn and Lehmkuhl, 2000). Equilibrium shorelines are also recognized as a dynamic state in which the geometry of a beach reflects a balance between materials, processes and energy levels (climate) that is most closely approximated by sandy beaches, but the idea of an equilibrium profile is widely utilized and can be simulated (Woodroffe et al., 2011: 422).

A 'dynamic' as opposed to an 'historical' approach to geomorphology emerged and was well in evidence by 1960. Hack (1960) advocated dynamic equilibrium as a reasonable basis for the interpretation of topographic forms in an erosionally graded landscape. Every slope and stream channel in an erosional system was seen as adjusted to every other, and when the topography is in equilibrium and erosional energy remains the same, all its elements are downwasting at the same rate. Hack interpreted the accordant summits of the ridge and valley province of the USA as the result of a dynamic equilibrium amongst ongoing processes rather than such surfaces being the remnants of earlier erosion cycles. However, these two apparently opposing views could be reconciled (Richards and Clifford, 2011) as it had been shown that in Maryland the Piedmont upland summits had been variously interpreted as a peneplain, a series of peneplains, a surface of marine planation, and a landscape in dynamic equilibrium, whereas these different perspectives are actually related to different scales of time and space (Costa and Cleaves, 1984). An excellent reconciliation of the timeless dynamic equilibrium and of time bound approaches was provided by a recognition of different timescales (see Chapter 14) of steady, graded and cyclic time (Schumm and Lichty, 1965). The subsequent debate was fuelled by the many definitions available (Thorn and Welford,1994) so that Kennedy (1992) questioned whether these displayed the growth of knowledge and understanding or rather was a sign of desperate scrabbling to put everything under one umbrella. If it were the latter it could justify 'requiem for a dead concept' (Kennedy, 1994) and asking whether the concept of equilibrium was truly central to geomorphology.

6.2 Development and application of related concepts

With the adoption of the systems approach (Chapter 2) it had already become accepted that balance or equilibrium was achieved between system components but that equilibrium is a highly ambiguous state reflected in the variety of definitions available to represent different types of equilibrium (Chorley and Kennedy, 1971: 201). Furthermore, many types of equilibrium purporting to describe geomorphic systems (such as dynamic and graded equilibrium) are seen to be very different from one another and, in turn, may also be different from the kinds of equilibrium resulting from broader systems analysis (Mayer, 1992). However, equilibrium concepts are unavoidable in any discussion of disturbances and responses in geomorphology (Phillips, 2011a), and notions of equilibrium influence not just the analysis of geomorphic responses but also perceptions and value judgments. The equilibrium types that evolved were not unique to geomorphology but reflected developments in other sciences: for example in ecology, where the classical view had assumed a state of equilibrium and stability, encapsulated in the 'balance of nature' paradigm. Most recently ecological systems have dominantly been interpreted by a non-equilibrium view, emphasizing the role of chance events such as disturbance (Perry, 2002).

As the numerous definitions are overlapping and sometimes contradictory, Howard (1988: 49–50) suggested the following definition and characteristics:

Equilibrium refers to a type of temporal relationship between one or more external variables or inputs and a single internal variable, or output, that has the following characteristics:

1. *Changes in the inputs must cause measureable changes in the output either immediately or after a finite time. This eliminates the trivial case of inputs that have no effect on the output.*

2. *The value of the output at a given time is related by a single-valued, temporally invariant functional relationship to the value(s) of the input(s) at the same time, within a consensual degree of accuracy.*

3. *The functional relationship should be capable of repeatable testing, either experimentally or observationally.*

4. *An equilibrium relationship may be limited to certain ranges of the input values and/or to certain rates of change of the input values.*

The most frequently used terms included in Table 6.3 develop from dynamic equilibrium, stability, and the steady state concept which Phillips (2009) suggests to be inextricably tied to a 'balance of nature' perspective. This postulates normative, self-maintaining states, thence leading to other terms including **quasi equilibrium**. The range of terms developed reflects the need to allow for different timescales (see Chapter 13) and for the changes and perturbations that may take place.

An excellent way of visualizing the alteration of equilibrium or stable positions was provided (Graf, 1977; 1988) through the application of the rate law in geomorphology. Graf (1977; 1988) represented change from steady state or equilibrium A to steady state D passing through reaction time B (the time needed for the system to absorb the impact of the disruption) and relaxation time C (the time period during which the system adjusts to new conditions) (see Chapter 14). Using this as a framework it is possible to illustrate several of the equilibrium types (Table 6.3) diagrammatically (Figure 6.1), showing how relatively stable steady, dynamic, or quasi equilibrium states can be altered to one of dynamic metastable equilibrium and how, following disruption, there can be adjustment to a new equilibrium condition.

6.3 The present status

Equilibrium remains an important concept in geomorphology but is no longer regarded as universal. As in ecology the 'balance of nature'

Table 6.3 Types of equilibrium

Development	Definition	Citation
Dynamic equilibrium	Fluctuations are balanced about a constantly-changing system condition which has a trajectory of unrepeated 'average' states through time.	Gilbert, 1877 Hack, 1960
Stability	Frequently used in ecology, concerned with the ability of a system to remain near an equilibrium point or to return to it after a disturbance.	Hill, 1987
Steady state	Input and output and properties of a system remain constant over time, allowing the modelling of complex natural systems.	Mackin, 1948 Howard, 1982
Steady state equilibrium	Constant form, not necessarily static.	
Dynamic metastable equilibrium	A condition of dynamic equilibrium with thresholds allowing occasional large fluctuations to produce jumps in form and process, initiating new dynamic equilibrium regimes.	Schumm, 1973; 1977
Quasi-equilibrium	A state of near equilibrium in the apparent balance between force and resistance, can be alternative term for steady state equilibrium, also dynamic equilibrium.	Langbein and Leopold, 1964
Metastable equilibrium	Occurs after adjusting to a threshold value condition in which stable equilibrium obtains only in the absence of a suitable catalyst, trigger or minimum force which carries the system state over some threshold into a new equilibrium regime.	Schumm, 1973; 1977
Non-equilibrium	Non-equilibrium view emphasizes the role of chance events such as disturbance has become dominant in ecological dynamics.	Perry, 2002
Equilibrium, disequilibrium, and non-equilibrium landforms	Equilibrium is a constant relation between input and output or form, toward which a landform tends or around which it fluctuates in time. Disequilibrium landforms are those that tend toward equilibrium but have not had sufficient time to reach such a condition. Some landforms, called non-equilibrium, do not tend toward equilibrium even with relatively long periods of environmental stability, but instead undergo frequent and large change in form. Landscapes contain mixtures of equilibrium, disequilibrium and non-equilibrium landforms.	Renwick, 1992

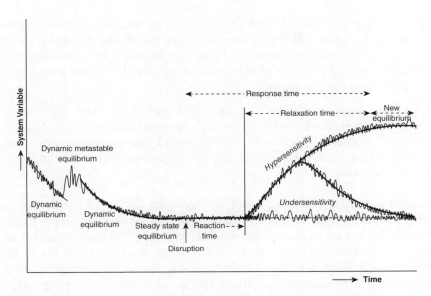

Figure 6.1 Types of equilibrium (based on Figure 3 in Gregory, 2005, and Gregory and Downs, 2008)

concept used to describe natural systems as being in a state of equilibrium, in which disturbing one element disturbs the entire system, has now become less accepted (Zimmerman and Kuddington, 2007; Kricher, 2009). Recent work on complexity in geomorphic systems has shown that there are no goal functions in Earth surface systems and no single normative condition for landforms and landscapes (Phillips, 2009) so that the idea of a single equilibrium condition has been succeeded by **multiple equilbria**, and then by **disequilbrium**. Such philosophical perspectives contrast with assumptions that are often made for practical purposes in which it is assumed that forms (such as river channel dimensions) are equilibrated or in balance with controlling variables in the short term.

In a paper entitled 'The End of Equilibrium' (Phillips, 1992) it was argued that although there is this pervasive concern with the presence and nature of geomorphic equilibria, the focus should now rest not on the detection of a single stable equilibrium condition for geomorphic systems but on the behaviour of systems away from equilibrium, the presence and prevalence of dis- and non-equilibrium forms, and the presence of multiple equilibria. Three categories of equilibrium concept can now be recognized (Phillips, 2011b):

1. *Relaxation time* equilibrium merely implies that a landform or landscape has completed its response to a disturbance; this is the weakest concept of equilibrium, implying only that a response has had time to be completed.

2. *Characteristic form* is a stronger notion of relaxation time equilibrium, with the additional criterion that the system achieves (or at least clearly moves toward) a form or state, such as a strongly concave longitudinal river profile, which is adjusted to its environmental constraints and context.

3. *Steady-state* equilibrium, the strongest and most restrictive variety, implies that characteristic forms and states, for example river channels which continually adjust to maintain steady-state sediment transport, are stable in response to all but the largest perturbations and self-maintaining.

Further discussion (Bracken and Wainwright, 2006: 173) questioned whether the general concept of equilibrium continues to support geomorphological investigation, or whether geomorphological equilibrium is an out of date and overcomplicated concept and just one of the metaphors widely used in geomorphology. Just as a new meta-language may be needed in biogeography (Stott, 1998), so it should be in geomorphology. In European and North American geomorphological research a viewpoint may have prevailed that landscape is inherently stable: the implicit assumption of equilibrium (or tendency towards it) has encouraged an approach that is inherently static and linear. Even where the variability from equilibrium is appreciated, geomorphologists have nevertheless sought straightforward correlations between landscape variables (Bracken and Wainwright, 2006: 176).

6.4 Conclusion

Simple notions of equilibrium may now be outmoded. They developed centrally within Euro-American research and may have led to inappropriate designs or landscape modifications (Bracken and Wainwright, 2006: 176). Perhaps the description '*A good servant but a bad master*' that Francis Bacon (1561–1626) applied to money, but which has also been applied to fire, can equally be applied to equilibrium. The concept remains relevant because non-equilibrium states still cannot be understood without an appreciation of equilibrium notions. From this platform of understanding, complex non-linear dynamical systems and non-equilibrium states will now be considered in the next chapter.

FURTHER READING

Bracken, L.J. and Wainwright, J. (2006) Geomorphological equilibrium: myth and metaphor, *Transactions of the Institute of British Geographers, 31*: 167–78.

Howard, A.D. (1988) Equilibrium models in geomorphology. In M.G. Anderson (ed.), *Modelling Geomophological Systems*. New York: Wiley. pp. 49–72.

Kennedy, B.A. (1994) Requiem for a dead concept, *Annals Association American Geographers*, 84: 702–5.

Mayer, L. (1992) Some comments on equilibrium concepts and geomorphic systems, *Geomorphology*, 5: 277–95.

Perry, G.W.L. (2002) Landscapes, space and equilibrium: shifting viewpoints, *Progress in Physical Geography*, 26: 339–59.

Phillips, J.D. (1992) The end of equilibrium? In J.D. Phillips and W.H. Renwick (eds), *Geomorphic Systems*. Amsterdam: Elsevier (and in *Geomorphology*, 5: 195–201).

Renwick, W.H. (1992) Equilibrium, disequilibrium, and nonequilibrium landforms in the landscape, *Geomorphology*, 5: 265–76.

Thorn, C.E. and Welford, M.R. (1994) The equilibrium concept in geomorphology, *Annals of the Association of American Geographers*, 84: 666–96.

TOPICS

1. Explore whether 'Despite the usefulness of equilibrium as a metaphor for understanding the complexity of geomorphic systems, the myth that has arisen around the concept has seriously undermined the practice of the discipline' (Bracken and Wainwright, 2006: 176).

 WEBSITE

For this chapter the accompanying website **study.sagepub.com/ gregoryandlewin** includes Tables 6.1, 6.2; and useful articles in *Progress in Physical Geography*. References for this chapter are included in the reference list on the website.

7

COMPLEXITY AND NON-LINEAR DYNAMICAL SYSTEMS

Notions of a single equilibrium and stable state were followed by recognition of the existence of multiple stable and unstable states with nonlinearity recognized as common in geomorphology. Such recognition, and also that major changes could occur as a result of relatively minor shifts, gives complex behaviour outcomes that are not predictable in linear systems. Research on complexity theory and non-linear dynamical systems has included concepts involving chaos theory, dissipative structures, bifurcation and catastrophe theory, and fractal patterning, as well as instability, resilience theory, adaptive cycles, and uncertainty. Such promising concepts are still developing with multidisciplinary centres established to progress further research.

Complexity and complexity science, now widely acknowledged throughout the physical and social sciences, have also been explained to general audiences. For example, Neil Johnson in his (2007) book *Two's Company, Three is Complexity: A Simple Guide to the Science of All Sciences* suggests that complexity is 'the study of the phenomena which emerge from a collection of interacting objects'. This approach to science is expected to 'define the scientific agenda for the 21st century' (Waldrop, 1992) and is one that has been influential in fields as diverse as physics, cosmology, chemistry, climate research, biology, neuroscience, clinical medicine, management and economics (Arthur, 1999).

Complexity was listed as one of four concepts by Schumm (1977), was the first of 10 suggested by Coates (1981), and is widely recognized in geomorphology (see Chapter 1, **Table 1.5**). Themes presented in the final sections of each of the preceding chapters need to be assembled and integrated – is complexity the concept which enables this to be achieved? An answer is attempted by considering the questions that have arisen and the themes that require integration revealing increasing complexity (7.1), the progress achieved by modelling complexity in non-linear dynamical systems (7.2), and the present multidisciplinary position (7.3).

7.1 Increasing complexity

Two trends were very influential as concepts developed over more than 100 years (Chapters 1–6). First, quantification and technical developments incorporated statistical and mathematical methods. Progress in remote sensing, in computing and the development of geographical information systems, was reinforced with advances in microprocessors. Although we now take such advances for granted they collectively enabled the discipline to collect and analyse more data, to address some problems in a way not previously feasible, and to employ more advanced modelling methods, especially as the Quaternary timescales became both defined and refined. Second, following the adoption of a systems approach, there was an agreed need for greater liaison between process studies and research on landform evolution; two strands which had been separate when processes were investigated on very short timescales and landform developments were studied on very long timescales. However, the greater attention given to short-term change reinforced the need for knowledge of research on one theme to assist the other. Several spatial scales and timescales were recognized including the steady, graded and cyclic times (Schumm and Lichty, 1965) together with a way of combining time and space scales (Chapter 13). Concepts emerging from awareness of the need to investigate adjustments over relatively short periods of time included complex response, whereby the same outcomes can be achieved from different sequences of development, or several different outcomes from singular external drivers; and thresholds which are the boundary conditions or tipping points separating two distinct phases or equilibrium conditions (Schumm, 1979). These are equally relevant to long-term outcomes of landform evolutionary models. Schumm identified three types of response:

- *Extrinsic*, threshold condition at which the landform responds to an external influence: for example, the way in which a river channel cross-section may change abruptly from a single-thread to a braided channel following a large-storm event.

- *Intrinsic*, where a geomorphic response is caused by exceeding a threshold in an internal variable without the need for an external stimulus: for example, long-term weathering reducing the strength of hillslope materials until slope failure occurs, potentially providing a significant change in sediment production to an upland river channel.

- *Geomorphic*, in which an abrupt landform change occurs as a consequence of progressive intrinsic or extrinsic adjustments: for example, where progressive river bank erosion eventually causes meander cut-off which prompts further changes related to bed-level adjustments in the channel.

Such concepts were joined by others (Table 7.1). In the last two decades of the 20th century plate tectonics revolutionized thinking about the surface of the Earth at the level of macroscale geomorphology (Chapter 5). This had produced the shift in focus so that the three foci for geomorphology could be seen as geographical, interpreting morphology and processes; geophysical, concentrating upon the broad structural outlines (see Church, 2005; Summerfield, 2005b); and chronological, focused on the history of change (Chapter 5, pp. 53–54).

Applying systems thinking to any one of these approaches, or to an integrated approach, involved many variables. Geomorphic systems can exhibit complex, apparently random behaviour and patterns in both spatial and temporal domains, consequences of the cumulative impacts of assembled sets of individual process-response mechanisms, or of multiple controls over process-response relationships (Phillips, 1992). Previously many of the relationships between variables had been considered as linear, as when changes in one variable produce small or proportional responses in another. Variables may be expressed in logarithmic or other forms to produce straight-line relationships, but 'non-linear' in geomorphology is generally taken to mean that there are relationship jumps, complex relationships between variables, or alternative pathways that may be followed as control variables (such as energy availability) change in value. It became appreciated that a basket of nonlinearities exists in geomorphology, thus opening up possibilities of complex behaviour not recognisable in linear systems (Phillips, 2003). As a result of a threshold-dominated nature, it was suggested that geomorphic systems are typically 'jumpy'. For example, stream power and transportable sediment size may be linearly related, but naturally varied stream sediment size mixes may mean that nothing much may happen in transport terms until a threshold is reached at which the largest particles are forced to move and the rest of the particle sizes with them.

Long-term change can also progress along multiple historical pathways rather than proceeding towards some single equilibrium state or along a cyclic pattern (Phillips, 2006a). Thus multiple stable and unstable equilibria, which may be the basis for new models of geomorphological evolution, succeeded ideas of single equilibria and stable states. In an unstable system there would be no return to equilibrium and it may not be possible to predict successive behaviour (Thornes, 1983b). When analysing temperate palaeohydrology, shifts in a single parameter may produce a sequence of changes involving a transition from a single stable equilibrium to a bifurcation with two stable states, and ultimately to further bifurcations with many stable states (Thornes and Gregory, 1991: 535). Thus conceptual frameworks emphasizing single-path single-outcome trajectories of change were supplemented by multi-path multi-outcome ones, with a transition from the idea of normative standards

Table 7.1 Some concepts developed towards the end of the twentieth century

Theory	Definition	Examples of Use
Indeterminacy	Results open to two or more conflicting interpretations.	Simpson, 1963
Complex response	The same outcomes can be achieved from different sequences of development (see Chapter 15).	Schumm, 1973
Thresholds	The boundary conditions or tipping points separating two distinct phases or equilibrium conditions (see Chapter 15).	Schumm, 1973
Hysteresis	Term imported from study of magnetism applied to bivariate relationship in a looped form with a different value of the dependent variable according to whether the independent variable is increasing or decreasing.	Wood (1977) applied to relationship between suspended sediment concentration and discharge for specific events. (See also Phillips, 2003.)
Sensitivity	The likelihood of responding to slight changes, expressed as the ratio between the magnitude of adjustment and the magnitude of change in the stimulus causing the adjustment.	Brunsden and Thornes (1979). Different interpretations can be employed (Downs and Gregory, 1995).
Multiple stable and unstable equilibria	May be the basis for new models of geomorphological evolution, with no return to equilibrium, and may not be possible to predict successive behaviour (see Chapter 6).	Thornes, 1983
Nonlinear dynamical systems (NDS) theory	Shows that extreme complexity can arise due to the nonlinear dynamics and couplings of relatively simple systems, includes specific techniques and concepts such as chaos, dissipative structures, bifurcation and catastrophe theory, and fractals.	Phillips (1995b) recognized 10 modes (five stable and five chaotic) of topographic evolution.
Multiple causality	Acknowledged in ecology and geomorphology through the process–form interaction.	Stallins (1996) in biogeomorphology.

such as characteristic (steady-state) equilibrium, zonal, and mature forms to the recognition that some systems may have multiple potential characteristic or equilibrium forms – and that some may have no particular normative state at all (Phillips, 2009). A simple analogy used for an environmental system was a ball moving within a bowl, so that under a single equilibrium or stable condition it would gravitate towards the centre of the bowl. The alternative is of an unlimited plane or an upturned bowl over which the ball moves and may never find a single stable position.

Such conceptual insights were important, particularly in relation to management, as underlined by Schumm's 'seven reasons for uncertainty' of which one is complexity (Schumm, 1985). He also proposed three types of misperception of fluvial hazards (Schumm, 1994) which were that any change is not natural (stability), that change will not cease (instability), and that there was excessive response (change is always major). Such misperceptions can lead to litigation and may have been the reason for unnecessary engineering works. Also in relation to managing rivers, a one-size-fits-all is inadequate (Wohl, 2013) because complex, non-linear behaviour must be considered. More multidisciplinary research on the Earth's surface, requiring strong links between geomorphology and other disciplines – including human dynamics, biology, biochemistry, geochemistry, geology, hydrology, and atmospheric dynamics, including climate change – means that new quantitative techniques for characterizing landscapes and landscape change can move towards a renaissance in theory and modelling (Murray et al., 2009).

7.2 Modelling complexity in non-linear dynamical systems

The necessary theory and modelling can be provided under the umbrella of 'complexity', which encompasses several related theoretical approaches that originated in non-linear dynamics (Murray et al., 2009: 497). Whereas systems were first thought of as predictable and analysable, often associated with linear models, research results showing that change could occur as a consequence of relatively minor shifts produced more uncertainty than previously appreciated. With the progression from systems (see Chapter 2) to **systems of complex disorder** and then to **systems of complex order**, recognized as having a large number of components, simple analysis techniques could no longer be employed. The extension of many of the ideas developed up to 1999 was crystallized by Phillips (1999) into 11 principles (**Table 7.2**), prompting the question as to whether Earth's surface systems become more diverse over time, to which Phillips (1999: 143) suggested an answer of 'yes, no, or maybe' depending on spatial and/or temporal scale and historical contingency.

Understanding complexity benefits from the study of complex adaptive systems involving many apparently independent agents interacting with each other in ways that are not always totally predictable: the recognition of self-organization involving non-linearity and feedback loops in which small changes can have substantial effects; and where the whole is seen as more than the sum of its (reductionist) parts (Pearce and Merletti, 2006). Complexity research in a range of disciplines from the late 1980s focused on the self-organization of complex systems, and has provided insights which have the potential to explain the functioning and evolution of landscape systems (Favis-Mortlock and de Boer, 2003).

Whereas open systems research was classically concerned with linear relationships in systems near equilibrium, sometimes with rapid or unexpected changes triggered by small events, a new direction was provided by the concept of *deterministic chaos* (Lorenz, 1963). The implication in geomorphology was that system outputs (or responses) are not proportional to system inputs (or forcings and drivers). Whereas the emphasis in earth science had been largely concerned with *stochastic complexity*, concerned with chance statistical variations as in variations about mean values or deviations from a trend, *deterministic complexity*, expressed in the non-linear dynamics and couplings of relatively simple systems represented by relatively small equation systems, has also been recognized by geomorphologists (Phillips, 1992). This is based on *non-linear dynamical systems* (NDS) theory, which includes specific techniques and concepts such as chaos, dissipative structures, bifurcation and catastrophe theory, and **fractals**. Use of NDS concepts in understanding earth surface processes and landforms can be mapped onto existing and even traditional theoretical concepts in geomorphology. Sources of nonlinearity in geomorphic systems largely represent well-known geomorphic processes, controls and relationships that can be readily observed (Phillips, 2003). These include abrupt thresholds, for example in sediment transport, or changes of form as in channel pattern transformations. Non-linear frameworks are necessary to explain some phenomena not otherwise explained: they can provide better analytical tools, improve the interpretation of historical evidence and usefully inform modelling, experimental design, landscape management and environmental policy.

An illustration is provided by the discriminant function between braided and meandering channels in which (**Figure 7.1a**) a channel at point B is stable whereas a channel at point A is potentially unstable because it is close to the threshold line. In practice, thresholds may be fuzzy with the overlap of pattern types (Lewin and Brewer, 2001; Eaton et al., 2010; Kleinhans and van den Berg, 2011). This may be because factors like bank stability are not included. Catastrophe theory, a branch of bifurcation theory (Thom, 1975) used as a basis for theory building (Graf, 1988), also provides an alternative to thresholds by recognizing zones of transition where two states of equilibrium are possible. The mathematical formulations of catastrophe theory can be used to account for sudden shifts of a system from one state to another, as a result of the system being moved across a threshold condition. Introduced to signify sudden changes or jumps that occur after relatively smooth progress, Rene Thom (1923 –) suggested seven types. The *cusp catastrophe* has been most used in geomorphology to model how small changes in parameters affecting a non-linear system can cause equilibria to appear or disappear, or to switch from repelling to attracting, giving rise to large changes in the behaviour of the system. A specific example (**Figure 7.1b**) illustrates a situation where the cusp catastrophe is applied to various channel patterns by relating the factors of stream power and resistance to the responding variable of sinuosity.

Because many environmental systems can be described as chaotic, producing responses which do not achieve a fixed equilibrium condition or value, chaos theory has been used to explain the fact that complex and unpredictable results can and will occur in systems that are sensitive to their initial conditions (Phillips, 2006b). Chaos theory studies the behaviour of dynamical systems highly sensitive to initial conditions, popularly referred to as the butterfly effect, showing how rich, complicated and perpetually dynamic behaviour can arise from simple non-linear interactions. Recent geomorphological research is revealing many cases where local deterministic interactions in a spatially distributed system can explain complicated behaviours that would previously have been ascribed to complicated (usually unknown) causes that defy holistic understanding (Murray et al., 2009).

Dissipative structures and fractals included in non-linear dynamical systems (NDS) theory show that extreme complexity can arise due to the non-linear dynamics and also the coupling together of relatively simple systems represented by relatively small equation systems (deterministic complexity), although virtually all the basic conceptual underpinnings of NDS theory can be mapped onto existing and even traditional theoretical concepts in geomorphology (Phillips, 1992). A typology has been proposed (Phillips, 2003) including thresholds, storage effects, saturation and depletion, self-reinforcing feedback, self-limiting processes, competitive feedbacks, multiple modes of adjustment, self-organization and hysteresis. Dissipative structures – a term proposed by the Russian-Belgian physical chemist Ilya Prigogine (1917–2003) who received the Nobel Prize for Chemistry in 1977 for his pioneering work on these structures – are thermodynamically open systems operating out of thermodynamic equilibrium. Cyclones and hurricanes are examples but there are also applications in geomorphology (**Table 7.3**). There is a close association between deterministic chaos and fractals, a term coined in 1975 by Benoit Mandelbrot (1924–2010), for the complexity of patterns which can be subdivided into parts, each of which is nearly a reduced-size copy of the whole. Coastlines, river networks, fault patterns, and the land surface as a whole (**Table 7.3**) have been analysed in terms of their fractal characteristics showing that, as higher resolution measurements are made, there is a rate of increase in length which is the fractal dimension.

Following on from the adoption of fractals and deterministic chaos has been the use of self-organizing concepts for complex systems, a field developed in thermodynamics, cybernetics and computer modelling. The first multidisciplinary conference on self-organizing systems, held in Chicago in 1959, prompted coordinated research particularly from the late 1980s. Self-organization is the spontaneous creation of an overall coherent pattern out of local interactions, typically with non-linear dynamics because of circular or feedback relations between the components (Heylighen, 2001), but without any necessary control by the environment or an encompassing or otherwise external system. Characteristics of self-organizing systems are feedbacks, complexity, scale differentiation and emergence (Favis-Mortlock, 2013). Negative feedback occurs when the system functions so that the disturbance effects are counteracted over time allowing the system to return to its pre-disturbance state, whereas positive feedback occurs when a disturbance forces the system away from its earlier state. Complexity, not precisely defined, characterizes a system with properties that cannot be described fully in terms of the properties of its parts.

Emergence relates to those structures that emerge as a consequence of the interactions between system components. Again, this cannot be simply inferred from the behaviour of the components, conforming to Aristotle's description in 330 BC of synergy as 'The whole is more than the sum of its parts'. A continuous flow of matter and energy through thermodynamically open systems, or dissipative systems, allows the system to maintain itself in a state that is far from thermodynamic equilibrium. Models are required to study self-organizing systems and one that has been employed is the cellular automaton (CA) approach which models continuous space into a series of cells, usually as part of a regular square or rectangular grid. This approach has become an archetypal model for several geomorphic systems common in many branches of geomorphic science (Fonstad, 2013). Used in relation to river channel processes and their geomorphic evolution, these physically-based approaches enable models to be applied to a range of spatial scales (1–100 km^2) and time periods (1–100 years) relevant to management (Coulthard et al., 2007). Such modelling may in one sense be deterministic, with invariant runs that are entirely dependent on initial conditions and input parameters, and they may then depend on linear sub-models:

in another way, individual catchment configurations may combine to produce patterning outcomes in discharge or sediment yield that other input forcing factors (like rainfall) could not anticipate. Alluvial systems are typically characterized by self-organizing complexity, as when components like meander development and cutoff processes interact to produce steady values for channel sinuosity (Stølum, 1996).

The study of self-organizing systems is particularly pertinent when environmental change occurs. Multiple possibilities could follow a disruption – uncertainties could arise from either extreme events or from unknown complications. The resistance of the system to change, its resilience, can be described as the capacity to absorb disturbance, and to reorganize while undergoing change but still retaining essentially the same function, structure, identity, and feedbacks (**Table 7.3**). In ecology, progress has been made towards a theory of adaptive change (Gunderson and Holling, 2001: 5) and the adaptive cycle comprises a forward and backward loop in four phases: *rapid growth* (r), *conservation* (K), *release* (Ω) and *reorganization* (α). In the forward loop, systems self-organize through rapid growth in which processes exploit and accumulate free energy (r) towards a point of maximum conservation and connectedness (K) epitomized by complex or mature states, such as forest ecosystems. In geomorphic terms, this process may be viewed as the evolution of landforms to a point of incipient instability where intrinsic or extrinsic thresholds are more easily exceeded. These include the concept that systems may exist in multiple steady states, flipping from one state to another as thresholds are transgressed. Panarchy is a conceptual term given to a nested set of adaptive cycles that cross multiple spatial and temporal scales (**Table 7.3**): it focuses the need to understand different scales of change in order to explain the causation of modern states.

7.3 Multidisciplinary complexity

Complexity therefore includes a range of concepts (**Table 7.3**) and is actively being explored in several disciplines, particularly in multidisciplinary research; several universities and other institutions have established centres for complexity analysis. Brad Werner, who authored an important paper on complexity in natural landform patterns (Werner, 1995), is now based in the Complex Systems Laboratory at the Cecil and Ida Green Institute of Geophysics and Planetary Physics, University of California, San Diego, investigating Earth and planetary surficial features and processes. A series of papers on landforms including aeolian dunes, beach cusps and patterned ground have been produced (see http://complex-systems.ucsd.edu/).

In the last twenty-five years, as it has become apparent that classical reductionist explanations may no longer be sufficient, it has become necessary to visualize suddenly changing oscillatory behaviour with systems switching between stable states (see Thornes, 2009: 135). Such non-linear frameworks can explain some phenomena not otherwise understood, provide better or more appropriate analytical tools, improve the interpretation of historical evidence, and usefully inform modelling, experimental design, landscape management and environmental policy (Phillips, 2003). This means that in recent geomorphological research many cases have been identified where local deterministic interactions in a spatially distributed system can explain complicated behaviour that would previously have been ascribed to complicated (usually unknown) causes that defied holistic understanding (Murray et al., 2009).

Not all questions have been answered by any means, so that uncertainty (of required parameters and predictions), sensitivity and indeterminacy are included in Tables 7.1 and 7.3. However, greater links between landform dynamics and landform change have been achieved to further illuminate understanding of the landscape. The sinking of the fishing boat the *Andrea Gail* with the loss of six sword fishermen in 1991 was as the result of a unique weather event created by so rare a combination of factors that meteorologists deemed it 'the perfect storm' (see Junger, 1997). This has subsequently been employed as a general metaphor for many situations to express the improbable coincidence of several different forces or factors to produce an unusual outcome. Thus Phillips (2007) conceptualized the 'perfect landscape' as a result of the combined interacting effects of multiple environmental controls and forcings to produce an outcome that is highly improbable. Geomorphic systems have multiple environmental controls, forcings and degrees of freedom in responding to them, allowing for many possible landscapes and system states so that, according to Philips (2007), the probability of existence of any given landscape is vanishingly small.

Greater links between human and physical systems have progressed with the modelling of human-landscape coupled systems (Werner and McNamara, 2007) and multidisciplinary research in complexity centres: we must retain open minds because non-linear systems are not always complex, and complexity does not always arise from non-linear dynamics (Phillips, 2005a, 2005b).

7.4 Conclusion

Concepts of complexity and non-linear dynamical behaviour are still in the early stages of application in geomorphology. They are being employed with some success in particular instances where higher-level combinations of process domains are involved. Linear behaviour is still commonly assumed as a working principle at lower level relationships between form elements and processes. Studies do suggest, however, that geomorphological predictions are difficult, not because of some kind of professional ignorance, but because large geomorphological systems are now understood to be intrinsically difficult to deal with in predictive terms because of the conceptual understandings outlined in this chapter.

FURTHER READING

Favis-Mortlock, D. (2013) Systems and complexity in geomorphology, *Treatise on Geomorphology,* 1: 257–70.

Murray, B., Lazarus, E., Ashton, A., Baas, A., Coco, G., Coulthard, T., Fondstad, M., Haff, P., McNamara, D., Paola, C., Pelletier, J. and Rheinhardt, L. (2009) Geomorphology, complexity, and the emerging science of the Earth's surface, *Geomorphology,* 103: 496–505.

Phillips, J.D. (2003) Sources of nonlinearity and complexity in geomorphic systems, *Progress in Physical Geography,* 27: 1–23.

Thornes, J.B. (2009) Time: change and stability in environmental systems. In N.J. Clifford, S.L. Holloway, S.P. Rice and G. Valentine (eds), *Key Concepts in Geography.* London: Sage. pp. 119–39.

Werner, B.T. (1995) Complexity in natural landform patterns, *Science,* 284: 102–4.

Wohl, E. (2013) The complexity of the real world in the context of the field tradition geomorphology, *Geomorphology,* 200: 50–8.

TOPICS

1. Explore the potential applications of the typology suggested by Phillips in *PIPG* (2003).

 WEBSITE

For this chapter the accompanying website **study.sagepub.com/ gregoryandlewin** includes Figure 7.1; Tables 7.2, 7.3; and useful articles in *Progress in Physical Geography*. References for this chapter are included in the reference list on the website.

SECTION B

SYSTEM FUNCTIONING

8
CYCLES

Cycles represent natural systems in which matter and energy are continuously transferred between different spheres of the environment. Their study involves the recognition of stores, fluxes and residence times, with hydrological, geological, and biogeochemical cycles providing the global context for landform science. Whereas cycles can be repeated, portions of cycles, as cascades or trajectories, are uni-directional. Temporal cycles include those in Milanković theory, as well as short-term cycles that enable different orders of magnitude of change to be identified, as cycles, or portions of cycles, that are more specifically geomorphic. Spatial cycles, usually referred to as cascades, involve erosion and deposition aspects of denudation, and geomorphology is central to understanding the temporal phasing of environmental flows.

A **cycle** is a natural process in which matter and energy are continuously cycled between different spheres of the environment (e.g., atmosphere, hydrosphere, geosphere, pedosphere, biosphere: see Gregory, 2010: Table 3.1). Any cycle contains stores, fluxes which are movements between the stores (the amount of material transferred from one store to another per unit time), and residence times (the amount of time spent in a particular store). Whereas a cycle consists of a set of events that can be repeated, portions of cycles – often referred to as cascades or trajectories – are uni-directional. In the context of system functioning we ask what is involved (cycles), later going on to consider why transport takes place (Chapter 9), how it occurs and when it takes place (Chapter 10), and with what consequences (Chapter 11). Each question is based on conceptual understanding. Some conceptual cycles, such as the Davisian cycle of erosion, have already been considered (Chapter 5), and three types of cycles are discussed here: global (8.1), temporal (8.2) and spatial (8.3).

8.1 Global cycles

Understanding of the functioning of global cycles emerged gradually in the environmental sciences with the major geological and hydrological cycles being appreciated for at least one hundred and fifty years, whereas understanding of biogeochemical cycles has been achieved more recently. The geological cycle is subdivided into the tectonic cycle and the rock cycle, although some geologists would also include hydrological and bio-geochemical cycles as integral to the geologic cycle. This demonstrates how cycles are fundamental concepts for all earth and environmental sciences. The tectonic cycle, also known as the orogenic or the geosyn-clinal cycle, embraces the production of ocean basins, continents and mountains. This was given great impetus by the advent of plate tecton-ics: the Wilson cycle, named after John Tuzo Wilson (1908–1993), refers to the cyclical opening and closing of ocean basins caused by move-ment of the Earth's plates, involving the formation of supercontinents (Rodinia, Pangaea) and their break-up over hundreds of millions of years (**Box 8.1**). In the lithosphere, which is c.100km thick, this involves the movement of large plates with junctions which can be divergent, conver-gent or transform. Plate tectonics provided a unifying concept for geology comparable to the influence of the evolution of species for biology by Darwin (1809–1882).

The **rock cycle** includes processes that are collectively responsible for the creation of igneous, sedimentary and metamorphic rocks, is affected by the hydrological and biogeochemical cycles, and is thought to have been originated by the ideas of James Hutton (1795) as part of his concept of uniformitarianism (Chapter 3). With the publication (Wilson, 1967) of plate tectonics, and what came to be known as the Wilson cycle, it was realized that plate tectonics could be the major driving force for the rock cycle. Many very graphic depictions of the tectonic and rock cycles, col-lectively making up the geological cycle, are available (e.g., http://www.geolsoc.org.uk/ks3/gsl/education/resources/rockcycle.html).

The **hydrological cycle**, also known as the water cycle, refers to the continuous movement of water on, above and below the surface of the Earth, involving major spheres of atmosphere and hydrosphere, and major processes of evaporation, condensation, precipitation, infiltra-tion and runoff. Many representations of the cycle are available and the USGS version is very graphic (**Figure 8.1**). Early philosophers had ideas about some aspects of the hydrological cycle, so that Marcus Vitruvius (c. 80–70 BC, died c. 15 BC), a Roman author, architect and engineer, described the way in which precipitation falling in the moun-tains infiltrated the Earth's surface leading to streams and springs in the lowlands, but it was Leonardo da Vinci (1500: 1452–1519) who recog-nized, on the basis of field observations, that river water derives from precipitation. The modern foundations of the cycle were established in 1580 by Bernard Palissy, a French potter (1509–1589, who envisaged the theory of the hydrologic cycle, and Pierre Perrault (1608–1680), whose òbservations of rainfall and streamflow in the Seine River basin confirmed earlier thoughts that rivers could be generated by precipitation thereby initiating modern scientific hydrology. Modern understanding of the hydrological cycle includes the ways in which water moves between stores represented by the oceans (c. 94%), groundwater (>4%), ice sheets and glaciers (c. 2%); with smaller amounts in the atmosphere and on land i.e., in lakes, rivers and soil. Solar energy drives the hydrological cycle, providing the lift against gravity required as water vapour is transferred by evaporation to form atmospheric moisture and then precipitation. The energy required to drive the global hydrological cycle is equivalent to the output of about 40 million major (1000MW) power stations (Walling, 1977), and about one third of the solar energy reaching Earth is used evaporating about 400,000 km^3 of water each year. Water (H_2O) that is constantly moving between the stores changes in phase from gaseous to liquid to solid in different parts of the cycle: such changes or fluxes are extremely important and complex but some, such as precipitation, occur for a relatively small proportion of the time.

Although the residence time that water stays in the stores of the atmosphere, the soil or the land surface is relatively short, it can be several thousand years in the oceans, groundwater or ice sheets, and in some arid areas (e.g., Algeria or Libya) groundwater is as much as 20,000–30,000 years old. The entire contents of the oceans would take about one million years to pass through the water cycle. Water volumes in the different stores have changed in the past, and some ten thousand years ago during the last ice advances there was more than twice as much ice over the Earth's surface (approximately 72 x10^6km^3 compared with 33x 10^6km^3 today).

Biogeochemical cycling – the cycling of life elements (C, O, N, S, P, H) affected by a combination of biological, chemical and geological processes, and involving atmosphere, oceans, sediments and living organisms – was probably envisioned after the development of biogeochemistry by Vladimir Vernadsky (1863–1945). This Russian scientist, who was instrumental in conceiving the noosphere (Chapter 16), embraced *The Biosphere* in a (1926) book in which he hypothesized how life processes shaped the Earth. These ideas were absorbed by the British scientist James Lovelock who suggested that the whole planet acted as a self-regulating and living entity, requiring the interaction of the totality of physical, chemical and biological processes to retain the conditions vital for the survival of all life on Earth, in turn controlling the atmospheric conditions required for the biosphere. Named the *Gaia Hypothesis*, after the Greek goddess of the Earth, this was formulated in the 1960s, and although largely ignored until the 1970s, it has subsequently become accepted as the Gaia theory. In 2006 Lovelock argued that change is now so substantial that Gaia may no longer be able to adjust, drawing the analogy of an old lady sharing her house with a growing and destructive group of teenagers – Gaia grows angry and if they do not mend their ways she will evict them (Lovelock, 2006: 47). The life elements include the carbon, oxygen, nitrogen, sulphur and phosphorus cycles, and their circulation can each be expressed diagrammatically. Other elements can also be visualized in terms of biogeochemical cycles including mercury (Hg).

The three major cycles – geological, hydrological and biogeochemical– together with their components, require specification. **Table 8.1** compiles information about the stores or reservoirs in each, and also about the typical fluxes and budgets. Other major cycles include **life cycles**, involving all the different generations of a species succeeding each other by means of reproduction, whether asexual or sexual.

8.2 Temporal cycles

Global cycles are spatial in that they apply universally at the present time. There are also a number of temporal cycles, considered more fully in Section C (see Chapter 13). Glaciation produced some of the most dramatic changes on the surface of the Earth when large continental ice-sheets developed. Five ice ages are known in Earth history, with the Quaternary the most recent, and each glaciation can be viewed as a glacial cycle, including successions of glacial and interglacials (in the case of the Quaternary lasting between two to three million years). The four major glaciations originally proposed for the Alps of Europe (Penck and Bruckner, 1901–1909), and for Scandinavia, Britain and North America, provided the framework for study of the Quaternary. Refined dating techniques, including the use of varves in lake deposits in Sweden since 1912, fossil pollen from peat deposits prompting palynology after 1916, and radiocarbon dating (^{14}C) applied to wood, charcoal, peat, organic mud and calcium carbonate in bones since 1949, contributed to the progressive refinement of glacial history. In the last three decades of the 20th century these techniques were supplemented by enormous advances in the identification of past timescales. Deep sea cores, dated by radiometric and other means, furnished an uninterrupted stratigraphical record for the last 900,000 years, and ice core evidence from Greenland and Antarctica established that the Quaternary began 2.8 million years (Ma) ago (Chapter 13).

Major fluctuations of this kind were ascribed to effects of the Earth's movements upon its climate by Milutin Milanković (1879–1958), a Serbian geophysicist and astronomer who proposed variations in the eccentricity of the Earth's orbit (a 96,000 year cycle), axial tilt or obliquity (cycles of c. 40,000 years), and the precession of the equinoxes (a 21,000 year cycle). Milanković's theory contends that variations in solar radiation/

energy received by the Earth occur in cycles which affect past climates and climatic patterns on Earth, with the three cycles (**Figure 8.2**) combining to affect the amount of solar heat that is incident on the Earth's surface, and subsequently influencing climatic patterns (for an explanation see www.sciencecourseware.org/eec/GlobalWarming/Tutorials/Milankovitch/). Work by Milanković in the first half of the 20th century was not fully appreciated until the 1970s when knowledge of deep-ocean cores and a seminal paper by Hays, Imbrie and Shackleton (1976) concluded that changes in the Earth's orbital geometry were the fundamental cause of the succession of Quaternary ice ages, and that a model of future climate predicts that the long-term trend over the next seven thousand years is toward extensive Northern Hemisphere glaciation. Although Milanković cycles are broadly established (**Figure 8.1**), theoretical difficulties have subsequently been discussed: for example, it has been suggested that the 100 kyr glacial cycle is caused not by eccentricity but by a previously ignored parameter, namely the inclination in the tilt of the Earth's orbital plane (Muller and MacDonald, 1995).

Short-term cycles have also been recognized, including two types of climate change during the last glaciation. Dansgaard-Oeschger cycles (D-O events) are rapid climate fluctuations that occurred 24 times, for 500–3,000 years, and consisted of an abrupt warming of up to 7°C within a few decades to near-interglacial conditions, followed by a gradual cooling. Each Dansgaard–Oeschger (D–O) event comprised a warm interstadial and a cold stadial (Williams, 2011). Related to some of the coldest intervals between D-O events were six distinctive events named Heinrich events, recorded in North Atlantic marine sediments as layers with a large amount of coarse-grained and land-derived sediments. These extreme short duration events represent the climate effects of massive surges of sediment-transporting icebergs into the North Atlantic.

Comparable climate cyclicity during the Holocene – referred to as Bond events and identified primarily from fluctuations in ice-rafted debris – may be the interglacial relatives of the glacial Dansgaard – Oeschger events. Eight Bond events with 1,470-year climate cycles in the Holocene have been identified, mainly based on petrologic tracers of drift ice in the North Atlantic.

Such temporal variations provide a context for geomorphological research enabling different orders of magnitude of change to be identified, including (Brown, 1991):

- major 'cyclic' time, including Milanković cycles (100ka, 44ka and 23ka), causing variations in climate involving large fluctuations in the water balance; such changes from glacial to interglacial climate conditions occurred more than 20 times during the Quaternary;

- climatic variations such as interstadials with a periodicity of either 40,000 or 25,000 years or less;

- change induced by climate fluctuations over periods of decades or centuries, instigating geomorphological adjustments (e.g., earth movement in Japan, Turkey or New Zealand) or human activity (e.g., deforestation);

- effects of individual events that may persist for many years, can cause thresholds (see Chapter 13) to be crossed to new states, and are increasingly 'readable' from the sedimentary record. A major flood can have long-term effects upon the landscape or lead to a threshold being crossed in order to move to a new state.

Timescales are explored further in Chapter 12 and concepts associated with temporal change in Section C.

8.3 Spatial cascades

Cycles at global and temporal scales that provide the concepts underlying much of landform science are equally basic to other sciences such as sedimentary geology, but some cycles, or portions of cycles, are more specifically geomorphological. Sometimes distinguished from the rock cycle, the **sediment cycle** involves the weathering of an existing rock, followed by the erosion of minerals, their transport and deposition, and then burial: parts of this cycle, usually referred to as cascades, are integral to such systems (Chapter 2). The solar energy cascade prompts geomorphic cascades, with hillslope-hydrological and nutrient cycle cascades (Chorley et al., 1984: 43–4) leading to denudation as shown in the sediment cascade. Denudation, erosion and deposition – perhaps some of the most fundamental geomorphic concepts – are not always consistently defined (Table 8.2). Sediment production, transport and deposition are central to landform investigations so that the sediment cycle, or more correctly the sediment cascade, is essential in geomorphology. As with global cycles, stores or reservoirs, typical fluxes and budgets can be identified, requiring an appreciation of familiar concepts that are basic to landscape dynamics (Table 8.2).

Sediment can be produced by endogenetic or exogenetic processes. Geomorphological aspects of sediments and solutes relate to the production, transfer, storage and deposition, thereby involving storage concepts (Table 8.2) and estimates of rates of denudation. Sediment production by erosion processes (Table 8.2) is succeeded by transport that can frequently be interrupted, leading to sediment storage. The terms buffers,

Table 8.2 Some conceptual terms associated with sediment cascades

Term	Definition
Denudation	Literally 'to strip bare' from the Latin *denudare*, broader than erosion because includes the results of all weathering, erosion and transport processes responsible for degradation of the Earth's surface
Erosion	The processes (see Table 9.2) whereby debris or rock material is physically loosened or chemically dissolved to be removed from a part of the Earth's surface
Deposition	Laying down of material accumulated after erosion and transport of material, after decay of organisms, or as a result of evaporation
Exogenetic processes	Occur at or near the surface of the Earth (see Section 9.2)
Endogenetic processes	Originate within the earth (see Section 9.2)
Sediment storage mechanisms	Explain the discrepancy between erosion and deposition, can include concept of landform impediments, including buffers, barriers and blankets
Exhaustion effects	A progressive reduction or exhaustion in the availability of sediment for mobilization and transport, recognized particularly in relation to generation of suspended sediment in streams
Capacity	The maximum load that can be transported by a specific process at a specific location, usually applied to rivers and streams
Competence	The maximum particle size transportable by the flow at a particular location
Sediment delivery	The fraction of sediment eroded within a catchment will reach the basin outlet and be represented as sediment yield (Walling,1983)
Sediment budget	The difference between the sediment input to a given area and sediment output from that area over a specified amount of time; if the budget is negative erosion must have occurred whereas if the budget is positive there must have been deposition
Denudation rate	A measure of the rate of lowering of the land surface by erosion processes per unit time, expressed in mm.1000 year $^{-1}$; these rates can be referred to as Bubnoff units

barriers and blankets (Fryirs et al., 2007) connote the interruptions to sediment conveyance by limiting the connectivity between landscape compartments. Buffers restrict sediment delivery to channels, barriers inhibit sediment movement along channels, and blankets drape channel or floodplain surfaces, affecting the accessibility of sediment to entrainment (Fryirs et al., 2007), all potentially operating as a series of switches which turn on or off the processes of sediment delivery, thus determining the effective catchment area at any given time. This means that (dis)connectivity needs to be considered in the operation of the sediment cascade because buffers, barriers and blankets can interrupt the longitudinal, lateral and vertical linkages in catchments (Fryirs, 2013): in this way various parts of a catchment may be actively contributing sediment to the sediment cascade and can be switched on, or inactive and switched off. Also in the hillslope-fluvial system, exhaustion effects may obtain as the sediment supply is insufficient for the transporting capacity of the transporting medium. Capacity – as the maximum load that can be transported by a particular process at a specific location – is usually applied to rivers and streams, whereas competence refers to the maximum particle size transportable by the flow at a particular location. As knowledge of landscape-forming processes increased, attention was directed towards denudation rates (Table 8.2) to give an idea of the rate at which landforms could be changing, with maximum rates reported including values equivalent to 10,000 Bubnoff units (Table 8.2) in Taiwan (Li, 1976). River loads, used for calculating denudation rates, have limitations, including the amount of storage that occurs in basins, that the dissolved load may include non-denudation sources, that average rates disguise how actual rates vary substantially, and that human activity has had substantial effects (Walling, 2006). It is also necessary to place such rates in the context of rates of orogeny because Schumm (1963) demonstrated that modern rates of orogeny are about eight times greater than the average maximum denudation rate.

As greater attention was devoted to the mechanics of the geomorphic system (Chapter 5) it became necessary (Walling, 1983) to unlock the 'black box' of sediment delivery. As only a proportion of the material eroded within a catchment reaches the catchment outlet, more attention has to be given to the transfer of material and thence to the detail of sediment cascades. At least one edited volume (Burt and Allison, 2010) has focused on the transport of sediment through the fluvial landscape. The sediment cascade was employed as an important conceptual model (Caine, 1974) to construct an overview of alpine geomorphic processes, describing them as a series of sediment flux cascades. To understand the role of erosion and deposition in geomorphic systems the cascade was applied (Warburton, 2011) to slope and stream channel subsystems incorporating three subsystems: the geochemical, the fine sediment, and the coarse detritus (Caine and Swanson, 1989; Warburton, 2007). The progressive sequence of change throughout the fluvial system is embodied in systems of channel classification, and Warburton (2011) cites the classification proposed by Nevins (1965) which is included in a comprehensive compilation of such classifications (Downs and Gregory, 2004: Table 3.1). This was used to develop a channel classification framework (**Table 8.3**) involving seven nested scales and providing a hierarchical context for a particular stream or river, although 'every stream is likely to be individual' (Hynes, 1975).

Can the cascade concept be applied to other land-forming processes? This sediment cascade concept is explicitly geomorphological and can embrace hillslope as well as the fluvial system, as shown in the network structure of coarse sediment pathways in a central alpine catchment, using (Heckmann and Schwanghart, 2013) numerical simulation models for rockfall, debris flows, and hillslope and channel fluvial processes to establish a spatially explicit graph model of sediment sources, pathways and sinks. The interaction of sediment pathways forms sediment cascades, represented by paths in a graph model. This can be extended to paraglacial geomorphology where six paraglacial landsystems have been identified: rock slopes, drift-mantled slopes, glacier forelands, and alluvial, lacustrine and coastal systems, each containing a wide range of paraglacial landforms and sediment facies, collectively conceptualised as the storage components of an interrupted sediment cascade (Ballantyne, 2002a). The concept is also employed in a cascading approach to a periglacial mountain slope to assess dynamics within the coarse debris system (Müller et al., 2014) and embodied in glacial process systems.

8.4 Conclusion

Flows in cycles therefore provide a background not only to geomorphology but also to other earth and environmental sciences, although certain aspects are particularly pertinent to land form science. More broadly *environmental flows* are proposed as a set of ecological-based stream flow guidelines designed to inform sustainable water resource management, supporting healthy riverine habitats and providing sufficient water supply for society (Meitzen et al., 2013), because it is argued that geomorphological understanding is central to **environmental flows** as it is the interaction between flow, form and substrate that influences habitat type, condition, availability and biotic use across space and time. Cycles and concepts in subsequent chapters of this section interlock with temporal investigations (Section C), underlining the significance of events as pinpointed by Stephen Jay Gould (1988) in *Time's Arrow, Time's Cycle*:

At one end of this dichotomy – I shall call it time's arrow – history is an irreversible sequence of unrepeatable events. Each moment occupies its own distinct position in a temporal sequence, and all moments, considered in proper sequence, tell a story of linked events moving in a direction.

At the other end – I shall call it time's cycle – events have no meaning as distinct episodes with causal impact upon a contingent history. Fundamental states are immanent in time, always present and never changing. Apparent motions are parts of repeating cycles, and differences of the past will be realities of the future. Time has no direction.

FURTHER READING

Burt, T. and Allison, R. (eds) (2010) *Sediment Cascades: An Integrated Approach.* Chichester: Wiley.

Fryirs, K. (2013) (Dis)Connectivity in catchment sediment cascades: a fresh look at the sediment delivery problem, *Earth Surface Processes and Landforms,* 38: 30–46.

Gregory, K.J. (2010) *The Earth's Land Surface.* London: Sage. (See especially Chapter 3.)

Hays, J.D., Imbrie, J. and Shackleton, N.J. (1976) Variations in the Earth's orbit: pacemaker of the Ice Ages, *Science,* 194: 1121–32.

Meitzen, K.M., Doyle, M.W., Thoms, M.C. and Burns, C.E. (2013) Geomorphology within the interdisciplinary science of environmental flows, *Geomorphology,* 200: 143–54.

Slaymaker, H.O. and Spencer, T. (1998) *Physical Geography and Global Environmental Change.* Harlow: Longman.

Warburton, J. (2011) Sediment transport and deposition. In K.J. Gregory and A.S. Goudie (eds), *The SAGE Handbook of Geomorphology.* London: Sage. pp. 326–42.

TOPICS

1. Research the origin of the concept of denudation, and then explore whether it is defined consistently at the present time. Do you conclude that standardisation is required?

2. Extend the buffers, barriers and blankets introduced as concepts useful in understanding sediment storage and transport (Fryirs et al., 2007) to the other process systems listed in Table 8.2.

3. The sediment cascade concept is illustrated for catchments, hillslopes and paraglacial systems. How can other process systems be embraced within the sediment cascade concept?

 # WEBSITE

For this chapter the accompanying website **study.sagepub.com/ gregoryandlewin** includes Figure 8.1; Tables 8.1, 8.3; Box 8.1; and useful articles in *Progress in Physical Geography*. References for this chapter are included in the reference list on the website.

9

FORCE-RESISTANCE

Energy, both kinetic and potential, is derived from solar radiation, gravity, and the Earth's geothermal energy, and is recycled partly through physical and biochemical processes and by human impact. This produces forces that determine the character of major categories of Earth surface processes. Statistical mechanics and gravity-driven fluid forces provide the basis for the analysis of many surface processes, with theoretical understanding assisted by the development of models and the availability of quantitative methods. Ideally mathematical statements would describe a model based on physical principles applicable to all surface processes. Hitherto a hierarchy of models has been employed, many involving force, resistance, stress and strain, progressing to the production of geomorphic transport laws, with models including hybrid indices such as factor of safety.

Geomorphic processes are mechanisms for the movement or flux of material or mass with movement initiated when force overcomes resistance. Environmental forcing in a broader sense (Chapter 13) is now often used as a term to express the external forces that provide energy for environmental systems. It is not apparent when the 'forcing' term became common in geomorphology, and its earliest use was in physics, biology and ecology. The concepts involved in force-resistance progress move from a consideration of sources of energy (9.1), to the basis for geomorphic processes (9.2), and thence to expressions of force, stress and strain (9.3).

9.1 Sources of energy

Forces require sources of energy derived from solar radiation, gravity, and the Earth's internal heat, but are also provided by human impact (Chapter 16).

Radiative forcing is defined as the difference between radiant energy received by the Earth and energy radiated back to space. **Solar radiation** is the greatest driving force, being some 7,000 times greater than the heat flow from the interior of the Earth with some 10^{25} joules per year of solar energy received by the Earth's surface. The solar constant is the radiation received at the top of the atmosphere by a $1m^2$ surface normal to the solar beam, and on average this is 1400 watts m^{-2} (where a watt is a measure of power, or the rate of doing work, in joules per second). Mass balance, echoing the conservation of mass principle which states that matter cannot be created or destroyed, relates to the flows of energy and mass (e.g., water) from the Earth's surface to and from the atmosphere. Many internet illustrations, varying in complexity, are available, and a simple graphic one is http://earthguide.ucsd.edu/earthguide/diagrams/energybalance/ (produced in a partnership between the Scripps Institution of Oceanography, COSEE California and the San Diego Unified School District). Energy received by solar radiation is continually redistributed by motion in the atmosphere and oceans, and the energy balance can be regarded as an open system in an approximately steady state (Chapter 2). During the redistribution of energy, land surface systems are coupled dynamically with the atmosphere through the physical processes associated with fluxes of energy, water, biogeochemicals and sediments, embodied within cycles (Chapter 8). Such fluxes – often involving a series of cascading system responses as illustrated by river hydrology or by glaciers – interact with atmospheric elements of temperature, pressure, humidity, precipitation and wind. These elements which combine to make up weather are the basis for climate, often thought of as the totality of weather conditions at a place over a period of time. In general it is the climates (specifically temperature and its range, and the nature and availability of precipitation) that exert major influences on the geomorphic processes operating in particular zones of the land surface. Solar radiation effectively produces an energy driving force at a rate 4,000–5,000 times faster than all the other energy sources combined (Slaymaker and Spencer, 1998). Radiative forcing is defined by the IPCC (2001) as ' a measure of the influence a factor has in altering the balance of incoming and outgoing energy in the Earth-atmosphere system and is an index of the importance of the factor as a potential climate change mechanism'. This may be modified through anthropogenic effects (Chapter 16).

Gravity is the force imparted by the Earth to a mass which is at rest relative to the Earth. Newton's law of universal gravitation states that all masses are attracted to each other, but as the Earth is rotating there is also a centrifugal force exerted on mass. Hence gravity is a combination of true gravitational force and the centrifugal force. The standard acceleration due to gravity at 45° latitude at sea level is 9.80665m s^{-2}. Because of the existence of gravity every portion of the Earth's surface possesses elevation energy or potential energy – mgh – combining its mass (m), the vertical interval between it and the position to which it can move (h), usually height above sea level), and (g), the acceleration due to gravity which will vary slightly because of latitude (the spherical nature of the Earth) and height above sea level. Sediment transport involving fragmented rock by gravity-driven moving water and ice, coupled with wind-driven processes, is the major process effecting the mass transfer of material – the major mechanism for re-forming the surface of the Earth.

The Earth's internal heat, or **geothermal energy**, drives the tectonic system, which in turn modifies h, and thus potential energy, through differential uplift. Geothermal gradient is the increase of temperature with depth below the Earth's surface; an average increase of 1°C for each 28.6m, or 30.5°C for every km, although at some plate margins rates of 80°C per km have been discovered. The hot and mobile interior of the Earth induces tectonic activity: the resultant uplift may have been up to 3,000m in places in the last twenty to thirty million years, and associated with earthquakes releasing elastic wave energy and volcanic activity.

At molecular scales there is energy holding matter together which has been described as **atomic energy** or **chemical energy**, arising from the electromagnetic forces binding atoms into molecules and molecules into liquids and solids. This energy category is particularly significant when considering the weathering breakdown of materials in different temperature regimes and involving variable material geochemistries. Energy is expressed in joules (the energy expended or work done) in applying a force of one newton through a distance of one metre (one newton metre or N·m), and amounts of energy available from solar, gravitational, geothermal and anthropogenic sources show the relative importance of the different types.

Energy in a scientific sense has been used as a term since 1807: two accepted types are kinetic energy (a function of the movement of an object), first described by Gustave-Gaspard Coriolis (1792–1843) in 1829, and potential energy (a function of the position of an object), suggested by William Rankine (1820–1872) in 1853. Lord Kelvin (1824–1907) was responsible for the formulation of the first and second laws of thermodynamics: the first law, also known as the law of conservation of energy, is that energy cannot be created or destroyed, and the second law, that the total mass of a given system of objects is constant. This principle of conservation of mass involves **entropy**, often taken to be a measure of 'disorder' (Chapter 10), with the second law stating that the entropy of an isolated system never decreases, because isolated systems spontaneously evolve towards thermodynamic equilibrium – the state of maximum entropy. An interpretation of the third law is that systems which use energy best survive (Phillips, 1999b). Thermodynamics provides an important background for geomorphic concepts (**Table 9.1**) but has not been used as much as it could have been (see Chorley, 2000).

9.2 Force and resistance

Energy induces the forces that drive processes on the Earth's surface, with force defined as the product of mass and acceleration so that a mass of 1kg accelerated at $1m\ s^{-2}$ involves a force applied of one newton. Isaac Newton (1643–1727) formulated three laws of motion in 1687, and these became the foundation for statistical mechanics as well as the basis for understanding Earth's surface processes (**Table 9.2**).

Processes that occur as a result of forces have been categorized as exogenetic and endogenetic (Chapter 8) since Walther Penck (1888–1923) published *Die Morphologische Analyse* in 1924. The book was not widely available until it was published in translation in 1953 (Czech and Boswell, 1953). Exogenetic (Greek *exo* outside) processes are at or near the Earth's surface, whereas endogenetic (Greek *endo* inside) processes pertain to the Earth's interior. Resistance to forces operating is provided by lithology and geological structure, and by internally generated resistance that arises from friction between materials, their cohesion, and chemical weathering resistance. Processes involve a transfer, i.e., a flux, of energy or mass at or near the land surface. Every Earth surface process can be visualized in relation to forces that lead to stress with characteristics that provide resistance, so that the efficacy of processes at any location depends upon the extent to which, and how often, force exceeds resistance. Major categories of the Earth's surface processes identified (**Table 9.3**) can be associated with forces and the resistance encountered which may be provided by friction. Progress towards understanding how geomorphic processes operate was effectively achieved by Strahler (1952: 923), who wanted 'geomorphic processes to be treated as manifestations of various types of shear stresses, both gravitational and molecular, acting upon any type of earth material to produce the varieties of strain, or failure, which we recognize as the manifold processes of weathering, erosion, transportation and deposition'.

Hence it was not until after the middle of the 20th century that geomorphologists generally began to seek a basic theoretical understanding, aided by the import of ideas from physics and engineering, the development of models, and the availability of more quantitative methods – despite exceptional work by early pioneers such as G.K. Gilbert. Greatly influenced by the ideas of W.M. Davis, it proved much more attractive to theorize qualitatively about the Earth's surface processes than to study them in the laboratory. Although the concept of exogenetic and endogenetic forces had been enunciated by Penck (1924) with some appreciation of controls upon the mechanics of geomorphological processes, prior to 1960 geomorphological texts hardly referred to underlying forces and the mechanics of landscape processes. Nevertheless there were at least six significant antecedents for the growing interest in processes (Gregory, 2000), many of which are illustrated in **Table 9.4**.

Changes occurred in the second half of the 20th century for three main reasons: first, greater awareness of the need to focus on the mechanics of processes; second, the availability of quantitative techniques, modelling and systems thinking; and third, the growth of related disciplines such as hydrology. Concepts fundamental to other disciplines were imported into geomorphology. **Table 9.4** includes some of the key developments in books and research contributions that were seminal to the progress of thinking. Internal and external influences are not distinguished, but it is notable how there were key common developments in relation to fluvial, coastal, hillslope and aeolian branches of geomorphology, and the background details (**Table 9.4**) enable those wishing to see how the approach to processes evolved.

In the six decades since 1952 geomorphic processes were viewed as manifestations of mechanical stresses operating on Earth materials to produce various forms of strain, but the concept expanded to encompass non-linear behaviour and time-dependent dynamics (Chapter 7) with process metaphysics now offering an alternative conception placing process, rather than mechanics, in a position of ontological primacy (Rhoads, 2013).

9.3 Expressions of force, resistance, stress and strain and their assessment

Two major categories of shear stress affecting Earth materials, gravitational and molecular, were recognized by Strahler (1952). Gravitational stresses activate all downslope movements of matter, and hence include all mass movements, fluvial and glacial processes. Indirect gravitational stresses activate wave- and tide-induced currents and winds. Phenomena of gravitational shear stresses were subdivided according to the behaviour of rock, soil, ice, water and air as elastic or plastic solids and viscous fluids. The order of classification is generally that of decreasing internal resistance to shear and, secondarily, of laminar to turbulent flow. Implications arising from this paper had provided the foundation for systems (see Chapter 2), because Strahler (1952) contended that 'A fully dynamic approach requires analysis of geomorphic processes in terms of clearly defined open systems which tend to achieve steady states of operation and are self-regulatory to a large degree. Formulation of mathematical models, both by rational deduction and empirical analysis of observational data, to relate energy, mass, and time is the ultimate goal of the dynamic approach'.

In an ideal world, mathematical statements describing a physical model based on physical principles would have been devised to express force, resistance, stress and strain, and the relationships between them. However, despite considerable progress towards analysis of force and resistance (**Table 9.4**), no single approach has yet been developed that applies to the complete range of geomorphic processes (**Table 9.3**). Instead approaches have been devised for specific processes, often based upon physical laws, so that various measures – each embodying a concept – have been employed to represent force, resistance, stress and strain, and also to provide combined measures in hybrid indices. Examples are collated in **Table 9.5**. Models are the vehicle for such measures to be specified because modelling is the way in which geomorphology develops 'from a subjective catalogue of phenomena into a coherent and rational discipline' (Chorley, 1967: 90), and 'whilst every landform might appear to be unique when explored in detail, there are generalizations of different kinds that can be invoked as means of understanding, explaining or even predicting landscape form and process' (Odoni and Lane, 2011: 155).

One of the practical difficulties encountered by geomorphologists is the heterogeneity of Earth surface materials, applied forces and resistances. For example, a river bed is subjected to the force of gravity-driven flowing water, but it consists of a range of particle sizes packed together and with smaller ones sheltered by larger ones. Turbulent flow also involves eddies of fast and slow motion and thus complexity in the forces applied. Near-bed processes turn out to be both complex and difficult to observe directly. Vegetation plays an important role in resisting erosion, both of soil on hillslopes, along riverbanks and on floodplains. Direct observation of sub-glacial processes is even more difficult.

The consequence is that an exact calculation of force and resistance has often been replaced by semi-empirical relationships involving surrogate measures, such as mean flow velocity or average shear stress, together with representative measures for a range of bed materials including vegetation. For example, resistance to river flow may be represented by empirical coefficients such as Manning's n (and also its reciprocal, Strickler's K, or Chézy's C). Robert Manning (1816–1897) was an Irish engineer who developed what later (with some adjustments) became known as the Manning equation following earlier work by others, including P. Gauckler in France (see **Box 10.1**). Where average stream velocity (V) is not measured, it may be estimated using the channel hydraulic radius (R) – a measure of flow efficiency which is the ratio of channel cross-sectional area divided by the wetted perimeter of the channel, slope (S), and an empirical and composite estimate of 'roughness' (n) which may depend on sediment or grain roughness, form roughness from bars and channel sinuosity, and vegetation and similar factors:

$$V = R^{2/3} \, S^{\frac{1}{2}} / n$$

This practical basis for deriving resistance is based on field or laboratory calibration as much as being 'physics-based' on fundamental concepts of force and resistance. There are alternatives, and in hydraulics research the Darcy-Weisbach friction factor (f), a dimensionless coefficient also related to velocity and hydraulic radius, is usually preferred. Like the Manning equation, what has become known as the Darcy-Weisbach equation, named for H. Darcy (1803–1858) and J. Weisbach (1806–1871), has involved numerous researchers in fluid dynamics such as A. Chézy (1718–1798), O. Reynolds (1842–1912) and Th. Von Kármán (1861–1963). Geomorphological applications of this large area of hydraulics-related research, dating from the 18th century onwards but only made computationally more straightforward in recent decades, are discussed in Gregory and Goudie (2011) and may be approached in detail through hydraulics texts such as Kay (1998). The stresses, strains and dynamics of glacier flow and bed erosion are well covered by Benn and Evans (2010).

Progress shown in **Table 9.4** reflects how studies of process were advanced by a range of techniques (Table 1.3) of which field investigations, laboratory research and quantitative modelling provided the three main planks. These three approaches, which require observation and measurement, developed at different rates: many researchers tended to emphasize field research, and experiments were established to collect data for analysis to reveal the nature of geomorphological processes. Because field research was time-consuming, equipment-demanding, and necessarily confined to research over a few years, an alternative was to establish laboratory experiments where considerable progress was made employing wave tanks, flumes or wind tunnels, but the results were not easy to interpret because of scaling problems. For example, whilst physical models may in effect speed up field process operation to allow them to be directly observed, a size scaling-down may introduce complications because of shallow flows or fine sediment cohesion. The third approach employing quantitative methods, using either statistical or mathematical models, was more easily founded on physical principles or mechanics but demanded an oversimplification of the complexity of nature. Of these three approaches the traditional popularity of the field approach prompted the shrewd comment by Chorley (1978) that 'whenever anyone mentions theory to a geomorphologist, he instinctively reaches for his soil auger'. The three approaches remained somewhat distinct for many years with results that were not easily reconcilable. However, developments in the late 20th century in remote sensing greatly enhanced computing power for GIS and, not least, great advances in the availability of equipment meant that field monitoring could now be in real time and managed remotely. Observational timescales could in effect be extended using dating techniques to monitor actual, though in human terms slow, process rates. Results to benefit from the three approaches could now be more interrelated so that modelling was no longer merely common in geomorphology but pervasive (Wilcock and Iverson, 2003), with geomorphological research very often involving feedback between various types of model and between models and field observations (Hooke, 2003).

Several classifications of geomorphological models have been proposed: initially a conceptual model is specified; this may be followed by a physical model allowing a laboratory setting to show how certain geomorphic processes operate; then supported by an analytical model if a simplification of boundary conditions allows equations to be specified. Finally, numerical models allow more complicated boundary and initial conditions, and permit the solution of differential equations that cannot be integrated in closed form (Hooke, 2003). A hierarchy of models employed in geomorphology (Odoni and Lane, 2011) usefully distinguishes data models and theoretical models (**Figure 9. 1**) in a way which aligns with the 'model theoretic view' (MTV). This clearly distinguishes the way in which models are initially based upon observations from the real world or from theory about the real world.

Despite the significant progress achieved (**Figure 9.1**), understanding what drives the output and coping with the computational requirements necessary to explore model uncertainties (Odoni and Lane, 2011: 166) provide continuing challenges. A consequence is to use meta models or emulators which are essentially 'models of the models' and have been used successfully in other disciplines. Many models involve force, resistance, stress and strain, and illustrations of the ways in which they have been deployed are included in **Table 9.5**. However, two groups of models that have received much attention are geomorphic transport laws and those based upon hybrid indices such as Factor of Safety.

As geomorphic processes are mechanisms or groups of mechanisms for the transportation of debris, all processes are alike in that material begins to move when the forces involved become greater than the resistance. How often this occurs is the frequency of the process (Carson and Kirkby, 1972: 99). Geomorphic transport laws are fundamental. When forces become sufficient to create shear stresses large enough to overcome the resistance of materials, then sediment is released. Examples of reasons for increases in shear stress are illustrated in **Table 9.6** (see Varnes, 1978). The ultimate requirement in the study of process is to obtain information on the transfer of energy or mass or flux within the physical environment, and finally to express the change in landscape over a period $T = t_2 - t_1$ either by integration of fluxes

$$\int_{t1}^{t2} q_t \, dt$$

or by the net displacement of system parameters $(x_{t2} - x_{t1})$. This leads to the question of what transport laws have been embraced within geomorphology – what Church (2010) sees as the kernel of any landscape evolution model: he contends that these laws must be applications of the relevant principles of physics, incorporating criteria for the mobilization, transfer and deposition of earth materials of varying character, thus applying to the range of geomorphic processes (**Table 9.3**). However, such laws are a compromise between physics-based theory that requires extensive information about materials and their interactions, which may be hard to quantify across real landscapes, and rules-based approaches which cannot be tested directly but which may 'validate' models if the model outcomes match some expected or observed state (Dietrich et al., 2003).

Geomorphic **transport laws** (Dietrich and Perron, 2006: Table 1) can be developed as mathematical statements derived from physical principles to express the mass flux or erosion caused by one or more processes. Such laws can be parameterized from field measurements, tested in physical models, and then applied over geomorphologically significant spatial and temporal scales (Dietrich et al., 2003). Although we may not have achieved completely satisfactory views of specific laws, we have useful approximations for some of them, and there are important common features in the underlying physics (Church, 2010). Church (2010) contends that one future possibility is concerned 'to firmly ground the discipline in physical principles – to find the fundamental transport laws that govern the redistribution of earth materials on the surface of the planet; to properly reconcile the effects of those laws over large ranges of space and time; and to understand the complex nature of geomorphological processes and effects within the Earth system – seem to dictate a future in which geomorphology becomes increasingly a geophysical science'. He sees the alternative for geomorphology as becoming 'more preoccupied with issues such as the broader definition of the Earth system, environmental change in that system and the dominance of human agency' (see Chapters 2,14 and 17).

Hybrid indices have been devised as exemplified by the factor of safety, the ratio of resistance to force, which has been used in studying slope stability. The **factor of safety** (F) is the basis for a conventional engineering approach (Hansen, 1984) expressed as:

$$F = \frac{\Sigma \text{ mobilized resisting forces}}{\Sigma \text{ disturbing forces}}$$

This has provided some unification for the study of mass movement processes although variations in stability will usually arise because of local changes in any of the parameters in the slope stability system (Hansen, 1984). However, the factor of safety approach can be applied at different scales and this needs to include the effects of endogenous as well as exogenous processes (Petley, 2011). Other models have been developed so that a study examining 278 landslide initiation points in the Klanawa Watershed, located on Vancouver Island, British Columbia, Canada, investigated the possibility of enhancing landslide susceptibility modelling by integrating two physically-based landslide models – the Factor of Safety (*FS*) and the Shallow Stability model (SHALSTAB) – with traditional empirical–statistical methods, that utilize terrain attribute information derived from a digital elevation model, and land use characteristics related to forest harvesting (Goetz et al., 2011). The factor of safety approach has been used in landslide hazard analysis but can also be used in analysis of riverbank recession (e.g., Samadi et al., 2011). In many cases models have to be coupled with hydrological models as illustrated by high-resolution coupled hydrology – slope stability models (CHASM) to improve landslide stability assessments in areas, such as Hong Kong, which are subject to dynamic pore pressure regimes (Anderson, 2003).

9.4 Conclusion

Understanding of geomorphic processes ideally requires knowledge of the quantitative basis for the transformation and rearrangement of mass at the Earth's surface by weathering, erosion and deposition. This differential transfer of material is what leads to the generation of land surface forms. Processes may be framed in terms of mechanics and/or chemistry, and adapted for practical use and the space and timescales of interest, in the form of geomorphic transport functions. However, a further development has been to explore the amount of work done by geomorphic processes, considered together with power in the next chapter.

FURTHER READING

Church, M. (2010) The trajectory of geomorphology, *Progress in Physical Geography*, 34: 265.

Dietrich, W.E., Bellugi, D.G., Sklar, L.S., Stock, J.D. and Heimsath, A.M. (2003) Geomorphic transport laws for predicting landscape form and dynamics. In P.R. Wilcock and R.M. Iverson (eds), *Prediction in Geomorphology*. Geophysical Monograph Series, number 135. Washington, DC: American Geophysical Union. pp. 103–32.

Odoni, N.A. and Lane, S.N. (2011) The significance of models in geomorphology: from concepts to experiments. In K.J. Gregory and A.S. Goudie (eds), *The SAGE Handbook of Geomorphology*. London: Sage. pp. 154–73.

Roy, A.G. and Lemarre, H. (2011) Fluids, flows and fluxes in geomorphology. In K.J. Gregory and A.S. Goudie (eds), *The SAGE Handbook of Geomorphology*. London: Sage. pp. 310–25.

Strahler, A.N. (1952) Dynamic basis of geomorphology, *Bulletin Geological Society of America,* 63: 923–37.

Wilcock, P.R. and Iverson, R.M. (eds) (2003) *Prediction in Geomorphology,* Geophysical Monograph Series, number 135. Washington, DC: American Geophysical Union.

TOPICS

1. With the background of contributions utilizing force and resistance given in Table 9.4, consider whether studies of ice, wind and water have been equally well developed in geomorphology

 WEBSITE

For this chapter the accompanying website **study.sagepub.com/ gregoryandlewin** includes Figure 9.1; Tables 9.1, 9.2, 9.3, 9.4, 9.5, 9.6; and useful articles in *Progress in Physical Geography*. References for this chapter are included in the reference list on the website.

10

GEOMORPHIC WORK

The amount of work done by geomorphic processes is a key geomorphic concept which led to magnitude frequency concepts, followed by power as the rate of doing work. Associated concepts focused upon energy expenditure and usage, including efficiency, entropy, and least work principle, with maximum efficiency at one extreme and minimum power expenditure at the other. Progress of such ideas regarding the significance of energy expenditure developed in the 1960s and 1970s has been the subject for further attention in the 21st century, especially as applied to river systems.

When we use equipment and machines as part of our everyday lives we automatically think about their efficiency and operating cost, but can such concepts be applied to natural systems? Appreciating force/resistance concepts affords a key to understanding processes which employ energy and involve geomorphic work, introduced by Wolman and Miller (1960), and described in a paper by Rhoads and Thorn (2011) as one of the most influential ideas in modern geomorphology. Wolman and Miller's paper has been one of the 10 most cited papers in geomorphology (Doyle and Julian, 2005). It focused on the magnitude and frequency of forces in geomorphic processes and was succeeded by attempts to analyse the rate of doing work in the physical landscape with the attendant implications. Hence this chapter explains the sequence of impact of the concept by focusing on geomorphic work (10.1), magnitude and frequency (10.2), and power (10.3).

10.1 The concept of geomorphic work

The concept of geomorphic work, although basic to geomorphology, has not been utilized as extensively as might be expected. In SI units (**Box 10.1**) work is measured in joules (J), where one joule is defined as the work expended by a force of one newton applied through a distance of one metre. Hence work done is the product of force (F) and distance (d) moved so that

$$\text{Work done} = F.\, d$$

This also has a time dimension – the amount of work accomplished over a specified time period. The now classic paper by Wolman and Miller (1960) focused on **effective force** in landscape development, and argued that the amount of work done during different events is 'not necessarily synonymous with the relative importance of these events in forming a landscape or a particular feature of the landscape'. Their paper concentrated upon the significance of event frequency as well as magnitude in terms of 'work done', and also on the formation of specific features in the landscape. Wolman and Miller contended that the movement of sediment by air or water is essentially dependent upon shear stress which they described by the equation:

$$q = k\, (\tau - \tau_c)^{\,n}$$

Where q is the rate of transport, k is a constant related to the characteristics of the material transported, τ is the shear stress per unit area and τ_c is the critical or threshold shear stress required to move the material. Wolman and Miller (1960: Figure 1) depicted the situation in a classic figure (Figure 10.1a) and argued that, if the stress is normally distributed and continuous (as in Figure 10.1b) and if the quantity or rate of movement is related to some power of this stress, then the relation between stress and the product of frequency times rate of movement must attain a maximum. Hence the recurrence interval, or frequency, at which this maximum occurs is controlled by the relative rates of change of q with stress x, and of x with time. They demonstrated this maximum as c in Figure 10.1. It was emphasized that this generalization holds only if the applied stress exceeds a threshold value, because below such thresholds no work is done in moving material.

   This concept introduced by Wolman and Miller (1960) was important not only because it introduced 'work' and the notion of magnitude and frequency, but also because it emphasized the theoretical basis in mechanics. It also identified the potential significance of thresholds – a concept later developed by Schumm (1979) (see Chapter 13, Section 13.3). The concept was employed in 1964 (Leopold et al., 1964: Figure 3–23) in relation to sediment transport in streams, although it was noted that it could also be extended to the transport of sand and silt by wind. Although this approach initiated an important concept, it has also been noted (Sullivan and Lucas, 2007) that over time the abscissa (Figure 10.1) has often appeared as 'discharge', an observable surrogate rather than the

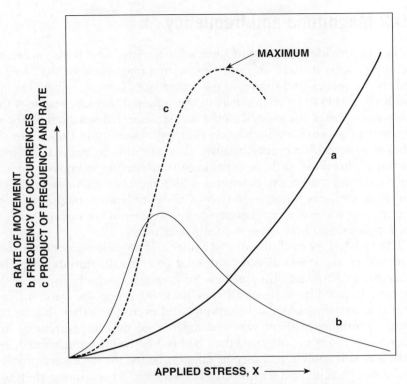

Figure 10.1 Relations between rate of transport, applied stress, and frequency of stress application (after Wolman and Miller, 1960, with permission of University of Chicago Press)

'applied stress' as originally published. However, such a change reasonably defines sediment transport in a channel only for flows up to bankfull depth because, once flows exceed bankfull, the rate of increase for in-channel bed shear (γRS) must show an inflection, with implications for calculations of total sediment transport at flows greater than bankfull.

And yet despite the progress towards adopting mechanics and physical principles (see **Table 9.4**) geomorphic work has not been applied as much as it could have been. This is surprising because the work done by geomorphic processes provides a fundamental and integrative index that should be fundamental in geomorphic analysis. The concept is important in geomorphology because the product of the relative frequency of an event and the magnitude of the event itself give an indication of the work done (Thornes and Brunsden, 1977: 9). A range of examples in **Table 10.1** illustrates specific ways in which geomorphic work has been employed in relation to processes in rainstorms, karst landscape evolution, the impact of landslides, solution on Alpine limestone cliffs, and work across floodplains.

10.2 Magnitude and frequency

Work can provide a measure of amount of activity but it is also necessary for any process to have an indication of the magnitude of that activity and the frequency and timing of its occurrence. Thus a review (Thomas and Kale, 2011) of river channels in the seasonal tropics concludes that geomorphic work is confined to the wet season; that seasonal rivers are adjusted to a range of discharges; that high-magnitude floods are significant geomorphic events because they are able to exceed the bank resistance threshold and are capable of transporting sediments (including coarse debris) on an enormous scale; and that inter-season fluvial transport rate can range over two to three orders of magnitude. This example shows how it is necessary to know when work occurs, what its magnitude is, and how often it occurs over time.

The 1960 paper by Wolman and Miller and the subsequent discussion (Leopold et al., 1964) directed thinking towards **magnitude-frequency concepts,** and towards the question of how often significant processes operate. It is salutary to recall that in 1960 processes were still not central to geomorphological thinking, and even when they did assume greater prominence there was a deficiency of data measurement and records. Methods of collecting data had to be invented and refined, and there was still some reluctance to take on board the necessary physical mechanics to achieve a depth of understanding. This context therefore explains why the Wolman and Miller (1960) paper, and the concepts it presented, were so significant. They were attempting to divine how frequent were the processes responsible for fashioning landforms and the physical landscape of the Earth's surface. This concept was introduced at a time when a more general consideration was being given to statistics of extreme values (e.g., Gumbel, 1958). It was concluded (Wolman and Miller, 1960), from analyses of the transport of sediment, that a large portion of work is performed by events of moderate magnitude which occur relatively frequently, rather than by rare events of unusual magnitude – a conclusion graphically demonstrated by the analogy they used (**Box 10.2**).

The broad conclusion reached in 1960 (Wolman and Miller, 1960) was that key formative events in many fluvial landscapes occurred about once every two to three years, but subsequently (Wolman and Gerson, 1978) this was amended to once in about every ten years. Such enquiries focused on ascertaining which events were responsible for particular features or characteristics of landscape. This therefore required an emphasis upon methods of analysis, including approaches involving recurrence intervals and probability analyses illustrated by flood frequency analysis; upon **formative events** described as 'those occurrences which shape the landscape and in some cases with the net denudation of specific meteorological events' (Wolman and Gerson, 1978: 190); and upon effectiveness, defined as 'the ability of an event or combination of events to affect the shape or form of the landscape' (Wolman and Gerson, 1978: 190). Subsequent enquiries demonstrated that it was unlikely that the same formative magnitude and frequencies would be obtained in all areas and be applicable to all processes, and that in some situations the impact of rare, catastrophic events could be more persistent and effective than in others (see Chapter 3), especially affected by the degree to which subsequent recovery is feasible. It was suggested (Wolman and Gerson, 1978: 205) that the frequency of formative events appears to be high in both periglacial and humid tropical regions and low in temperate or truly arid regions.

Over the last four decades research investigations have provided much greater knowledge about the magnitude and frequency of formative events, and key developments have included the relative significance of comparatively frequent and of rare catastrophic events, and the compilation of data on rates of erosion and denudation in different areas of the world. The concept that fairly frequent events accomplished the most significant work prompted direct observation and measurement in practical field programmes of a few years' duration. The results of decade-long field observations of mass movement processes on Karkevagge, Scandinavia, by Rapp (1960) quantified the relative importance of processes affecting a slope in a subarctic environment, showing that solution was the most effective agent of removal. In order to compare the work achieved by particular geomorphic processes, and then to compare these with other environmental fluxes, erosion intensity was suggested to be a measure of geomorphic work (Caine, 1976). Current investigations continue to address magnitude and frequency aspects. The potential of using a simple digital video camera technique to document the magnitude, frequency and geometry of grain flows on desert sand dunes has been demonstrated using data from lee side avalanches on a Namib crescentic dune (Breton et al., 2008). There have also been attempts to establish the upper limits of processes which affect the land surface. What are the greatest events ever recorded and what are the envelope curves that indicate their upper limits? Further studies have shown that different processes are maximized under different event frequencies–channel outlines to bankfull flows (which themselves recur with different frequency in different environments), but solute or sediment transport at somewhat lower flows. There has also been evidence that extreme events produce particular effects such as catastrophic landslides which may introduce mass possibly reworked subsequently by other processes in later more frequent events. This was not apparent in the relatively short-run data sets as analysed by Wolman and Miller. Examples of extreme events and their impact are provided in Table 3.2.

Recent progress has placed magnitude and frequency in the context of spatial and temporal scales including hierarchies (e.g., de Boer, 1992) in the context of the systems approach (see Chapter 2). The continuing questions for geomorphological research are how much work processes achieve, when is this work accomplished, and what are the events responsible? Recent research provides many examples of ways in which magnitude and frequency analyses are required (**Table 10.2**), including hillslope erosion by rainstorms, mass movement including debris flows and landslides, bankfull analysis, and sinkhole hazards. Such research underlines the need to analyse temporal variations that can be useful for spatial prediction (Galve et al., 2011) and for the study of climate change impacts (Korup et al., 2012).

10.3 Power

Further 'work' concepts that have been utilized include power, entropy, work minimization and efficiency.

Power, seen as the rate of doing work and overcoming resistance, offers an integrated theme (Gregory, 1987: 3) that embodies an essentially geomorphic concept that is potentially valuable for research analysis. Power expressed in watts, which are joules per second ($J\ s^{-1}$), was first introduced in relation to the rate of sediment transport. It was defined (Bagnold, 1960) as the product of fluid density (ρ), slope (s), acceleration due to gravity (g) and discharge (Q) in:

$$\omega = \rho Q g s$$

'Gross' or total **stream power** – the overall energy that a stream has available to do work, especially in the transport of its sediment load – is per unit length of channel (usually in metric units, as incorporated into Q, discharge in cubic metres per second). This may also be expressed as 'specific stream power' per unit width (w) of channel ($\rho Q g s / w$ or $\rho d v g s$, where d is depth and v velocity). In hydraulics and fluvial processes attempts to analyse the processes involved have used a number of variables, whereas expressing energy expenditure as stream power is more manageable analytically using field data that are quite commonly available. The potential energy that water possesses at a particular location is proportional to its height above some datum, which can be sea level or a lake level, and this potential energy is converted into kinetic energy as the water flows downhill under the influence of gravity. Three important aspects of stream power are how it is expressed, what controls it, and how it has been utilized and applied (**Table 10. 3**).

An approach by Andrews (1972) devised total **glacier power** (WT) as the product of basal shear stress and the average velocity. Effective power (WE) was determined by the proportion of the total average velocity which resulted from basal sliding so that the ratio between total glacier power and effective power (WT/WE) would vary according to the proportion of basal slip to internal ice deformation, thus enabling the distinction between arctic and temperate glaciers to be expressed. Andrews proposed that WT/WE is small – between zero and 0.2 for polar and subpolar glaciers, but between 0.5 and 0.8 tending to 1.0 for temperate glaciers – with the implication that glacial erosional forms produced by the two types of glacier differ in form and geometry. This shows how power can be employed for comparative analysis. Caine (1976) estimated the physical work in joules involved in different types of sediment movement, and also extended the comparisons to show how the range of geomorphic processes proceed at power densities ranging from 10^{-7} to 10^{-6} Wm^{-2} (**Figure 10.2**). Power can be used to compare the work of the largest known floods so that Baker and Costa (1987) compared power per unit area for flash floods ranging from 251 to18,582 Wm^{-2} whereas great historic and prehistoric river floods could be up to 2.5×10^5 Wm^{-2}. Power can also be used to compare geomorphic processes with other aspects of world energy flux (Gregory, 1987: 20–21; Smil, 1991). When comparing different situations it is usual to employ a unit measure, as illustrated by unit stream power defined as the time of potential energy expenditure per unit weight of water or width of an alluvial channel (Yang and Stall, 1974).

One of the concepts developed after 1960 has emphasized association and indeterminacy (Leopold and Langbein, 1963), where **indeterminacy** referred to those situations where the applicable physical laws could be satisfied by a large number of combinations of indeterminate variables. Energetic bases for explanations in geomorphology were still pursued, concerned with the energy available, its distribution, the amount used for specific processes and the optimum amounts that can be utilized. The aspiration was to discover the underlying principles that could account for variations in landforms or aspects of landforms according to the way in which energy is utilized. Thus in stream flow, energy may be consumed through turbulence and in overcoming frictional resistance, as well as in bed or bank erosion and sediment transport.

The series of concepts adduced (**Table 10.4**), many imported from thermodynamics, includes enthalpy which is a measure of the total energy of a thermodynamic system, and **entropy** which in thermodynamics provides a measure of molecular disorder and amount of wasted energy in a dynamical transformation from one state or form to another. Entropy, concerned with the distribution of energy and the probability of its distribution at a particular time, was introduced in

1850 by the German physicist R. Clausius (1822–1888) for the dissipative energy use in thermodynamics during a change of state. It was not imported to geomorphology until its use as an expression for the degree to which energy has become unable to perform work (Chorley, 1962) in a systems context with an application to landscape evolution explored by Leopold and Langbein (1962), and then to systems more broadly (Chorley and Kennedy, 1971). Several types of application have been identified in geomorphology as shown in **Table 10.5**. However, the use of entropy as an analogy has been criticized (Kennedy, 1985: 155) and found to be of dubious validity in fluvial geomorphology (Davy and Davies, 1979). Of several applications, two by Graf (1988) may be noted (**Table 10.5**). First, by analogy with the amount of heat liberated or absorbed by a perfect engine, the available energy in a system declines until the system decays to a state of no available energy and maximum entropy. An alternative application concerns the degree of system order and disorder – as the energy of a system is expended progressively less is available for work so that 'disorganization' is associated with high entropy values.

For effectiveness and rate of use, the concept of **efficiency** (**Table 10.4**), although usually thought of in relation to machines, can also be applied to natural systems by deriving power output/power input as a percentage. Although valuable in biology, for example in relation to the efficiency whereby plant tissue is assimilated by grazing animals, it is not as easy to extend to geomorphic systems, although maximum efficiency can obtain as a result of the optimum design of system components so that they work together most economically (Chorley and Kennedy, 1971: 225). Efficiency has been used in relation to the rainsplash process by comparing the change in potential energy of the particles with rainfall energy, demonstrating that the efficiency of the rainsplash process can be calculated at between 0.05 and 0.2% (Brandt and Thornes, 1987: 66), and also to show how sand transport efficiency is related by a power relationship to roughness element height (Gillies and Lancaster, 2013).

A variety of approaches to energy expenditure has been used to explain the characteristics of landform systems and these include minimum energy dissipation rate (mEDR) as a general theory applied to fluvial hydraulics (Yang, 1972). To allow for the effect of sediment transport, a subsequent development was to maximize sediment-transporting capacity (Pickup, 1977), shown (Kirkby, 1977) to be a possible fundamental principle of fluvial systems. A further analytical possibility is to use minimum stream power (Chang, 1979) which has been applied to river channel patterns. Such **extremal hypotheses** apply to alluvial river channels to some extent, but it has been suggested (Nanson and Huang, 2008: 927) that their selection has been somewhat arbitrary. Why should a river maximize or minimize any particular condition? An alternative

Least Action Principle (LAP) was advocated by Nanson and Huang (2007): introduced by Pierre-Louis Maupertuis in 1746, LAP may be used to demonstrate that, out of many possible alternatives, nature follows the path that is most 'economical' in terms of work. This concept has received strong support from a number of notable mathematical physicists, and it may provide an effective way of unifying subjects and consolidating theories in several branches of science in terms of the broad similarities between various areas of physics. For fluvial geomorphology this is argued (Nanson and Huang, 2007: 938) as achieving three important advances: first, it provides a convincing physical explanation of equilibrium states, which the LAP produces; second, the LAP presents a more straightforward and simpler object-orientated fuzzy method for solving river regime problems; and third, it demonstrates that river channel planforms are the product of the interactions between endogenous and exogenous factors, so that in fluvial systems with surplus energy this surplus can be expended on slope and/or channel geometry adjustments.

Applications of both extremal hypotheses and LAP have been largely used for research on fluvial systems, but there are possibilities for developing such concepts in other branches of landform science.

10.4 Conclusion

When processes were first investigated it was inevitable that the theoretical basis depended upon the concepts available, imported from either hydraulics and fluvial mechanics for understanding the dynamical basis of geomorphology (Strahler, 1952), or probability theory in the case of frequency analysis and the statistics of extremes. Great progress has been achieved during the last six decades but there is further potential to apply work, magnitude and frequency, power, and the LAP, to more sophisticated models of landform construction and change.

FURTHER READING

Gregory, K.J. (1987) The power of nature. In K.J. Gregory (ed.), *Energetics of Physical Environment: Energetic Approaches to Physical Geography.* Chichester: Wiley. pp. 1–31.

Korup, O., Görüm, T. and Hayakawa, Y. (2012) Without power? Landslide inventories in the face of climate change, *Earth Surface Processes and Landforms,* 37: 92–9.

Magilligan, F.J. (1992) Thresholds and the spatial variability of stream power during extreme floods, *Geomorphology,* 5: 373–90.

Nanson, G.C. and Huang, H.Q. (2008) Least action principle, equilibrium states, iterative adjustment and the stability of alluvial channels, *Earth Surface Processes and Landforms*, 33: 923–42.

Rhoads, B.L. and Thorn, C.E. (2011) The role and character of theory in geomorphology. In K.J. Gregory and A.S. Goudie (eds), *The SAGE Handbook of Geomorphology*. London: Sage. pp. 59–77.

Thompson, C. and Croke, J. (2013) Geomorphic effects, flood power, and channel competence of a catastrophic flood in confined and unconfined reaches of the upper Lockyer valley, southeast Queensland, Australia, *Geomorphology*, 197: 156–69.

Wolman, M.G. and Gerson, R.A. (1978) Relative scales of time and effectiveness of climate in watershed geomorphology, *Earth Surface Processes*, 3: 189–208.

Wolman, M.G. and Miller, J.P. (1960) Magnitude and frequency of forces in geomorphic processes, *Journal of Geology*, 68: 54–74.

TOPICS

1. LAP has been applied to fluvial systems (see Nanson and Huang, 2007). Suggest ways in which it is pertinent as regards applications to other geomorphic systems.

 WEBSITE

For this chapter the accompanying website **study.sagepub.com/greg oryandlewin** includes Figure 10.2; Tables 10.1, 10.2, 10.3, 10.4, 10.5; Boxes 10.1, 10.2; and useful articles in *Progress in Physical Geography*. References for this chapter are included in the reference list on the website.

11

PROCESS-FORM MODELS

The interaction of process and landform is central to geomorphic investigations and a series of concepts have been associated with the models of landscape development suggested over the last century. Process investigations were enhanced by considering the way in which specific landscape features are related to processes, as illustrated by grade, characteristic angles, drainage density and river channel capacity. Technique developments, especially of cosmogenic dating, have revitalized some earlier models. The complex response concept affords the reconciliation of alternative landscape histories, and a panoply of models is now becoming available offering opportunities to realize the objectives of the original qualitative approaches.

Investigations of Earth-forming processes inevitably led towards a consideration of whether there are landforms that are characteristic of particular areas or process regimes, raising questions about specific manifestations of processes expressed in the physical landscape. Answers to such questions were seen as the objective of some of the earliest models (11.1), with the advent of process investigations came the search for more specific features related to particular processes or to facets of those processes (11.2), and most recently ideas of multiple equilibria have shown that there are multiple possible outcomes (11.3), thereby combining several themes introduced in earlier chapters.

11.1 Landscape evolution models

From the end of the 19th to the middle of the 20th century geomorphology was dominated by models that relied upon a qualitative understanding of the relation between form and process and suggested the nature of progressive evolution. Indeed the very term process was seen as amounting to a progression through time, without any specification of underlying physical mechanisms. Although the detailed concepts now associated with temporal change are explained in Section C (pp. 125–178), the landscape models (Chapter 5, pp. 48–49), which were so influential prior to the understanding that arose from a greater appreciation of process, are outlined here.

The ideas of W.M. Davis (1850–1934) were a dominant influence for the first half of the 20th century, arising from more than the 500 papers and books that he produced. This inspired an 874-page second volume, within the series of four published on the history of the study of landforms, devoted almost exclusively to Davis (Chorley et al., 1973). He proposed a normal cycle of erosion to classify any landscape according to the stage it had reached in the erosion cycle, whether youthful, mature or old age, and provided a trilogy for understanding landscape in terms of structure, process and stage or time reached in an erosion cycle. This schema had biomorphic underpinnings not unrelated to evolutionary ideas dominant at the time. Whereas the normal cycle was the work of rain and rivers, arid landscapes were fashioned under the arid cycle of erosion, and in the marine cycle particular attention was given to shorelines of emergence or submergence. In addition there were two principal 'accidents' to the normal cycle, namely volcanic activity and the glacial accident. A striking feature of his work was the clarity of the supporting sketches and illustrations that assisted in the acceptance of his approach. Davis was seen as the prototype geomorphological 'fashion dude' by Sherman (1996: 107, 93).

Much has now been written about the dominance of the Davisian approach in the first half of the 20th century: major reactions, reasons for its success, and criticisms levelled are summarized in **Box 11.1**. It is arguable that geomorphology advanced more with the focus provided by the Davisian cycle than it would have done without it, especially as the approach of G.K. Gilbert (1843–1918), with a foundation of stream and landscape mechanics, could not be realized without the technical and quantitative developments that did not begin to arrive until the 1960s. Although there were sequels to the Davisian approach (see Gregory, 2000: 41) alternatives were also proposed, and the major ones are included in **Table 11.1**: an alternative suggested (posthumously in German in 1924) by W. Penck (1888–1923) was not widely known in the English-speaking world until its translation by Davis in 1932, and a more thorough translation in 1953 (Czech and Boswell, 1953) focused on slope replacement (**Table 11.1**). Rhoads and Thorn (2011) suggested that Penck saw possible pathways reflecting the changing interplay between uplift and erosional intensity (rates), his conceptualization so precise that Young (1972) was able to create a canonical diagram of it, in effect producing a process–response model. Lester King (**Table 11.1**) used parts of the Davis and Penck models, developing his model, first applied to South Africa, which involved the parallel retreat of slopes, ultimately producing a pediplain by the coalescence of pediments, and linked to large-scale crustal deformation which he described as cymatogeny. It is likely that his ideas were influenced by the particular nature of African landscapes, with extensive plains and scarps, whilst Penck was familiar with active tectonic regimes in South America.

Although acknowledging the ideas of Davis and Penck, L.C. Peltier (1950, 1974) related climate to geomorphological processes on a semiquantitative basis using mean annual temperature and mean annual precipitation. In identifying a periglacial cycle (**Table 11.1**) Peltier engendered criticisms similar to those advanced against the Davisian cycle of erosion, but his introduction of 'periglacial morphogenesis' stimulated a climatic geomorphology which was later emphasized by Tricart (1957) and Tricart and Cailleux (1965). Climatic geomorphology introduced a new concept into geomorphological teaching but did not readily become known in the English-speaking world until a paper by Holzner and Weaver (1965). In his approach Peltier produced a general landform equation (see also pp. 46–47) in which landform (LF) could be viewed as a function of geological material (m), rate of change of geological material, structural factor (dm/dt), rate of erosion (de/dt), rate of uplift (du/dt), and the total duration of the process (t) in the form LF = f (m, dm/dt, de/dt, du/dt, t). French (2011) has suggested that for cold environments the polygenetic Davisian model suggested by Peltier in 1950, involving congelifraction and leading to cryoplanation, remains the best.

Climatic geomorphology was also advanced by Julius Büdel (**Table 11.1**) who distinguished three generations of geomorphology:

1. Dynamic – concerns the study of particular processes.

2. Climatic – considers the total complex of processes in their climatic framework.

3. Climatogenetic – which involves the analysis of the entire relief including those features produced by contemporary climate and also those produced by former climates.

He proposed five climatogenetic zones (1963), later expanded to seven (Büdel, 1969) and then eight (Büdel, 1977), with the concept of the double etchplain developed for the tropics. As deep chemical weathering could be the norm in many tropical areas, creating a lower basal weathering front marking the position at which chemical weathering was attacking sound unweathered rock, there was also an upper surface of levelling where exogenous processes were eroding, transporting and depositing sediment across the land surface. The basal weathering front was composed of chemically weathered rock with occasional protrusions of unweathered residuals. The relative functional behaviour of these two levels (the 'double etchplain concept' reflecting contemporary and past erosion systems) could vary from humid tropics to semi-arid and arid landscapes. This model has been adopted in subsequent research (e.g., Thomas, 1978, 1994) and was incorporated in the climato-genetic geomorphology developed from climatic geomorphology by Büdel (1969, 1977), providing a way of visualizing the pattern of morphogenetic systems over the Earth's surface during the Cainozoic. This included the many temperate areas with tor-like residuals that were seen as remnants of times when climatic conditions were warmer and wetter, involving a double surface of levelling (Büdel, 1957). In fact Büdel did not use the term etchplain, a term first used by Wayland (1933), but it did arise from a translation of his work from German. This is one example of the way in which terms for erosion surfaces, including peneplains and pediplains, have evolved in papers by different authors as analysed by Ebert (2009).

A contrasting explanation, suggesting that dynamic equilibrium was a more reasonable basis for the interpretation of topographic forms in an erosionally graded landscape, was that every slope and stream channel in an erosional system is adjusted to every other, and when the topography is in equilibrium and erosional energy remains the same, all elements of the topography are down-wasting at the same rate (Hack, 1960). In this model, a rediscovered version of Gilbert's (1877) view of the interaction between form and process, accordant summits in areas like the ridge and valley province of the USA were interpreted as the inevitable result of dynamic equilibrium rather than as remnants of earlier erosion cycles. Although the principle of dynamic equilibrium was not an evolutionary model such as the Davisian geographic cycle, it can be used to explain specific landscape features and problems when it is assumed that the landscape has developed during a long period of continuous down-wasting (Hack, 1975).

Such ways of regarding the consequences of processes upon form in the physical landscape (although not all involved specification of physical processes and dating in any modern sense) were instrumental in the development of landform science. Such approaches, dedicated to explaining contemporary landform characteristics, were dominant in the first six decades of the 20th century, but their prominence was relinquished when processes with quantitative methods received greater attention. A resurgence of interest in these models has now become feasible as a result of the impact of plate tectonics and cosmogenic dating. Hence the most recent reviews (e.g., Bishop, 2011; Rhoads and Thorn, 2011) now show how concentrations of cosmogenic nuclides can provide estimates of the timing and rate of long-lasting geomorphic processes (Cockburn and Summerfield, 2004). Exposure age and erosion rate may be determined by analysing cosmogenic isotopes, provided the erosion rate is constant, thus advancing the understanding and interpretation of landscape development by dating and quantifying rates of landscape change over timescales from several thousands to several millions of years (**Table 11.2**). The significance of calculated erosion rates and exposure ages depends strongly on the models used to interpret isotopic data, the validity of assumptions inherent to these models, and the geologic surroundings in which the samples were collected (Bierman, 1994). Nevertheless, physical models geared towards understanding specific geomorphic processes, and numeric models specifying the coupling and feedbacks between Earth surface and geodynamic processes, are being developed (Pazzaglia, 2003): 'Since about 1990, the field of long-term landscape evolution has blossomed again, stimulated by the plate tectonics revolution and its re-forging of the link between tectonics and topography, and by the development of numerical models that explore the links between tectonic processes and surface processes', and with the recognition of denudational isostasy in driving rock upflift 'the broader geoscience communities are looking to geomorphologists to provide more detailed information on rates and processes of channel incision, as well as on catchment responses to such bedrock channel processes' (Bishop, 2007: 329).

11.2 Characteristic form

Whereas the models discussed so far (11.1) focused on the assemblage of landforms, albeit partial in some cases, at the scale of individual landforms there have been attempts to analyse sizes and shapes that characteristically develop in association with particular processes. Such enquiries address whether, in a particular situation, there is a general form to which all landforms tend and are there processes, or combinations of processes, that determine such characteristic forms? Although this concept has not often been explicitly stated, characteristic form (Chapter 6) is one of the current ways that equilibrium is recognized (Phillips, 2011a). It underlies much geomorphological thinking in the mid 20th century and Huggett (2011: 174) depicted interactions between form and process (**Figure 11.1**) as the core of geomorphic investigation. In citing examples it is not easy to be comprehensive, but grade, slopes, drainage density, subglacial bedforms, meander planforms and river channel capacity exemplify the approaches taken. This develops from ideas on landform (Chapter 4), equilibrium (Chapters 3 and 6) and complexity (Chapter 7), and provides a foundation for the elaboration of a complex response.

One of the most durable concepts has been that of **grade**; it was used by Gilbert and by Davis, stimulated many subsequent papers (e.g., Dury, 1966), and also stimulated an excellent survey by Chorley (2000). This has all the ingredients of a key and typical concept in geomorphology: origin in another discipline (engineering); attention given by early geomorphologists (G.K. Gilbert); absorption into a cyclic model (Davisian cycle); being questioned as a result of process investigations (Mackin, 1948); and culminating in a consolidation of disparate views for present understanding (some accepted and others fallen away). The concept has evolved through a series of phases (**Table 11.3**), reminding us that exponents of concepts were particular individuals. J. Hoover Mackin (1905–1968) was described (Chorley, 2000: 568) as someone who 'developed into the proverbial absent-minded professor who mislaid vital keys, forgot where he had parked his car, often wore mismatched socks (and sometimes shoes!) and, when travelling, left a trail of belongings ... His lectures consisted of a barrage of words which were delivered while he drew on the blackboard and chainsmoked ... He was a perfectionist who possessed great clarity of mind'. The sequence of phases through which the concept of grade evolved (**Table 11.3**) is prefaced by a simple definition and concluded by a summary of present interpretation; it is not much employed in contemporary research but it is necessary to understand the way in which such concepts have been shaped. A term like grade may in fact be used to represent different concepts and it is necessary to understand what each researcher has in mind.

Characteristic slope angles were identified in slope analysis (Young, 1961) as 'those which most frequently occur, either on all slopes, under particular conditions of rock or climate, or in a local area'. A quantitative analysis of frequency of occurrence of slope angles on 13 different types of rock outcrop in parts of Eskdale, Yorkshire (**Figure 4.2**) allowed the identification of characteristic slope angles for each lithology (Gregory and Brown, 1966). Limiting slope angles are those 'which describe the range of angle within which given forms occur or given processes operate, either on all slopes, under particular conditions of rock or climate, or in a local area' (Young, 1961; 1972: 165). A limiting angle on a particular lithology at a specific location is the angle just below that at which slope failure occurs by mass movement: the limiting angle indicates a threshold for slope stability, and hence may be related to the factor of safety. As the factor of safety involves the ratio of resistance to force (Chapter 9), if the factor is less than 1.0 the slope will fail through mass movement to achieve a lower and more stable slope angle. However, it is not easy to compute all the forces acting, calculate total resistance, or derive values in comparable units. Although useful at the stage when geomorphology was becoming more quantitative and exploring the potential of mapping methods, characteristic and limiting angles of slope are now appreciated to be area and condition dependent. From a modelling viewpoint Kirkby (1997: 128) contended that 'The *characteristic form* is one in which denudation at every point is proportional to its elevation above some base level ($-dz/dt \propto z$). Where the slope base itself is held at a fixed elevation, then this elevation is the base level. Characteristic forms also apply where the slope base itself also declines towards some base level elevation – behaviours closely analogous to a Davisian view of peneplanation'.

Two pertinent characteristics of the fluvial system are drainage density and river channel capacity. Whereas the drainage network is itself a concept, its dynamics have to be placed in the context of modes of water flow through the drainage basin (**Figure 11.2a**) and the quantitative measure most frequently used is drainage density (D_d), the total length of stream channels (ΣL) in the area of a drainage basin (A_d):

$$D_d = \Sigma L / A_d$$

Drainage density has been seen as an important characteristic of the drainage network since a focus on morphometric analysis followed from the classic paper by Horton (1945). It has been demonstrated that it can reflect different lithologies, aspects of climate and degrees of human activity, as well as significantly reflecting hydrological response from a basin because water flows more rapidly in channels than it does through other routes available (see **Box 11.2**). Drainage density can be an indicator of process and it also – through its influence upon flood routing – affects hydrograph generation, thereby proving useful in hydrological basin modelling. However, drainage density values have to be obtained carefully and interpreted in light of the fact that a drainage network is dynamic, embracing perennial, intermittent and ephemeral channels (**Figure 11.2 a**), so that the density that affects peak flows can be very different from the density that relates to low flows. Greater resolution by satellite imagery has encouraged revived use of drainage densities, but the care required for their interpretation in the light of their dynamic significance still pertains. Drainage density illustrates how a concept needs to be defined consistently with reference to the conditions obtaining when it was sampled, and the limitations of the information sources used.

The cross section of a river channel is another landform which can be related to process, and **river channel capacity** is a way of characterizing the dimensions of a river channel at a specific location (**Figure 11.2c**), enabling channels to be compared spatially throughout one drainage basin or between drainage basins. A definition is not straightforward: channels can be compound in cross section; they may not be clearly differentiated from the floodplain; and they may alter due to short-term storm events. Capacity is usually defined as the cross-sectional area of the river channel as far as the sharp morphological breaks in slope at the contact with the floodplain (Downs and Gregory, 2004). At any specific location along a river, river channel capacity reflects the interaction of water and sediment discharge with local factors including sediment in the bed and banks, vegetation and slope. The frequency of discharge at the bankfull stage, or the channel capacity discharge, was thought to be a 'dominant discharge' cor-responding broadly to the mean annual flood with a recurrence interval of 1.58 years on the annual flood frequency series (Gregory and Madew, 1982). However, subsequent research showed that the frequency of occur-rence of the bankfull discharge was a range of flows with recurrence inter-vals in the range one to ten years (Williams, 1978). Components of the drainage basin respond to dynamic conditions in a particular way. During different storm events, just as the river channel cross section and the asso-ciated floodplain have different degrees of inundation to specific levels, so these levels relate to the different stages of river channel pattern in a reach, and to the different stages of extension of the drainage network as indicated in **Figure11.2b** (Gregory, 1977).

Although early research concluded that dominant discharge with a par-ticular recurrence interval accounted for the cross-sectional area of the chan-nel, it was later realized that a range of flows controls channel landform (see Knighton, 1998), that the significant controlling discharges can vary along the course of any one river, and – in some areas – short-term sequences of events of differing magnitude (e.g., flood- or drought-dominated condi-tions) may affect channel morphology (**Box 11.3**). River discharges can be estimated from channel dimensions (Chapter 9) and such relations have been used in the United States (e.g., Osterkamp and Hedman, 1982) and in the UK described as the channel geometry technique (Wharton et al., 1989; Wharton, 1992, 1995).This a useful means of first estimation of discharge for ungauged river channels.

Whereas these examples refer to specific landforms, a broader problem (Rinaldo et al.,1994) is whether the topography of a particular landscape is in balance with current climate-driven processes or contains relict sig-natures of past climates – a long-standing question in geomorphology. Employing a mathematical model of landscape evolution, Rinaldo et al. found that both cases – contemporary balance and relict features – are possible. This shows how models (11.1) and characteristic form (11.2) are pertinent to temporal investigations, as shown in Chapter 15.

11.3 Multiple outcomes in landscape models

The four examples (11.2), illustrating how landforms are related to particular process and environment characteristics, are now of historic interest to show how specific process-response systems were analysed. A related concept, **geomorphic effectiveness** (Rhoads and Thorn, 2011: 69), has been used to connote how the impact of a formative event in shaping the landscape depends not only on how much a specific event reconfigures the landscape morphology but also on the capacity of post-event processes to 'undo' the morphological changes sustained. This has been applied to hydrologic events, where some conceptual confusion exists (Beven, 1981), but it allows comparison of major flood events, such as three very large 20th century floods of the Indian Peninsula (Kale, 2007), and the relative significance of a major jökulhlaup in south east Iceland (Magilligan et al., 2002). Landscape modification in some areas is accomplished only by large floods exceeding specific thresholds so that the geomorphic processes are effective only in the largest events, as in the case of the Missoula flood (Benito, 1997). This can be pertinent to hillslopes and mass movement processes so that, for landslides, the 'work peak' refers to a population of landslides doing the most work in a given landscape (Guthrie and Evans, 2007).

Although the drivers for models and for characteristic form reflect system adjustments and change over time (Section C, Chapters 12–16), some aspects relate to contemporary processes although bridging the gulf between space and time. Complex response was proposed by Schumm (1973) to designate how several alternative change scenarios might result from comparable environmental conditions. Although initially associated with crossing a threshold, this has subsequently been considered more generally, relating to the complexity considered in Chapter 7 where it was argued that conceptual frameworks emphasizing single path outcome trajectories of change have been supplemented by multi-path multi-outcome perspectives (Phillips, 2009). The crossing of thresholds can trigger spatial variations in response that are complex response – a distinctly non-equilibrium mode of system dynamics.

Models were originally constructed (11.1) assuming that processes were understood, or at least didn't need to be if process was taken to be simply progression in time, followed by a stage in which attention was given to specific landforms and the way in which they were related to processes (11.2). The present situation is that new models are rapidly developing, aided by developments in modelling computation, DEMs, GIS, new techniques (e.g., lidar, cosmogenic dating reviewed in 11.1 above), and conceptual developments. **Table 11.4** includes several examples of such recent research focused on process-landform or process-response systems, selecting just a few examples that relate to glacial, arid and coastal as well as fluvial systems. It is impossible to give comprehensive illustrations, but these examples demonstrate the nature of the recent advances that have revolutionized the way in which process-response systems can now be investigated. Several of these have implications for the study of environmental change. Process and form research necessarily involves research covering extended time periods, raising the question of when contemporary investigations become the temporal-based study of environmental change?

The end of this section (B) dealing with system functions emphasizes the need to relate the study of contemporary processes with those investigating changes over time (Section C): it underlines the need to link our understanding of the impact of catastrophic events (Chapter 3); of deciding where equilibrium assumptions are justified (Chapter 6); where concepts of complexity and non-linear analysis are appropriate (Chapter 7); and how force, power, form and process should be linked in a geomorphological context (see Phillips, 2013: 34). In a very different context William Faulkner, in his book *Requiem for a Nun*, included a very famous line – **'The past is never dead. It's not even past'.** Is this not an apposite text for concepts which link present functions with those relating to change over time?

FURTHER READING

Bishop, P. (2011) Landscape evolution and tectonics. In K.J. Gregory and A.S. Goudie (eds), *The SAGE Handbook of Geomorphology*. London: Sage. pp. 489–512.

Chorley, R.J. (2000) Classics in physical geography revisited. Mackin, J.H. 1948: Concept of the graded river. Geological Society of America Bulletin 59, 463–511, *Progress in Physical Geography,* 24: 563–78.

Guthrie, R.H. and Evans, S.G. (2007) Work, persistence, and formative events: the geomorphic impact of landslides, *Geomorphology,* 88: 266–75.

Hack, J.T. (1960) Interpretation of erosional topography in humid temperate regions, *American Journal of Science,* 258: 80–97.

Huggett, R.J. (2011) Process and form. In K.J. Gregory and A.S. Goudie (eds), *The SAGE Handbook of Geomorphology*. London: Sage. pp. 174–91.

Kirkby, M.J. (1997) Tectonics in geomorphological models. In D.R. Stoddart (ed.), *Process and Form in Geomorphology.* London: Routledge. pp. 121–44.

Martin, Y. and Church, M. (2004) Numerical modelling of landscape evolution: geomorphological perspectives, *Progress in Physical Geography,* 28: 317–39.

Phillips, J.D. (2014) State transitions in geomorphic responses to environmental change, *Geomorphology.*

TOPICS

1. Suggest other examples of characteristic form to add to those in Section 11.2.

2. Discover when base level was suggested as a concept and explore its evolution.

WEBSITE

For this chapter the accompanying website **study.sagepub.com/ gregoryandlewin** includes Figures 11.1, 11.2; Tables 11.1, 11.2, 11.3, 11.4; Boxes 11.1, 11.2, 11.3; and useful articles in *Progress in Physical Geography*. References for this chapter are included in the reference list on the website.

SECTION C

SYSTEM ADJUSTMENTS

12

TIMESCALES

The timescales adopted in geomorphology fall well within the c.4.6 billion years of Earth history, with some being a mere season or even a single event. In addition to continuous timescales, discrete periods of Earth history have been utilized. Six hierarchical levels are formally defined geologically, and these embrace the external or allogenic drivers for the long-term intrinsic or autogenic processes that have fashioned the Earth's surface, some parts of which still bear ancient traces, whereas others have been fashioned more recently or are currently active. Contemporary problems demand attention to recent timescales, the Quaternary and the Holocene, although these are less formally partitioned. Geomorphology-focused classifications have also been attempted with short, medium and long timescales conceived in relation to system states. An outstanding challenge is to reconcile research at one timescale with results from another.

Geomorphological systems can be set within two overlapping time frameworks: that set by external change and that by geomorphological processing itself. Land-forming processes operate in the context of an Earth history that, as far as we know at present, has had a unique trajectory amongst planetary bodies. Of underlying interest to geomorphology are particular stages like ocean formation, probably around 4 billion years ago, and not very long after the Earth itself formed some 4.6 billion years ago. This led to water circulation via the atmosphere, cryosphere and continental run-off. There then followed on-going process cycles and sequences (Chapter 8) that have determined the nature of many of the surface forms we now see. Also there has been a long-lasting tectonic history of mantle convection and crustal plate development and consumption, including mountain building, continental fracturing and amalgamation, vulcanism, and episodes when the oceans overlapped continental surfaces to greater or lesser extents. This created a range of timed *endogenetic* (or internally driven) potential energy situations through crustal uplift and depression (Summerfield, 1991; Bishop, 2007; Bull, 2007). These have interacted with the great variety of *exogenetic* (or surface) process regimes across the global surface at different time periods (Gregory, 2010).

There has also been the development of life forms, and in particular generations of vegetation covers across many land surfaces since the *Silurian Period* (c.470–430 million years ago) when a supercontinent, Gondwana, existed and there were high sea levels (Corenblit and Steiger, 2009). This biological envelope has varied as continental climates have fluctuated and ecosystems have evolved, with episodes of both warming and glaciation. The biological interface between physical processes and land surface characteristics is of considerable significance, for example in weathering, in promoting greater landform stability, and also in the operation of the carbon cycle following changes arising from climate modification. We are beginning to know a great deal about such episodes within the most recent *Quaternary Period* (the last 2.7 million years) (Bell and Walker, 2005; Woodward, 2014), but there were also earlier glacial and warm 'greenhouse' episodes in Earth's history. Human beings are a life form too, and there has been both a purposive and an inadvertent history of transformations wrought by human activity: deforestation, cultivation, soil erosion, urbanization, fossil fuel consumption, and the management of rivers for irrigation, water supply and flood control (Gregory, 1995). The timing and intensity of these anthropogenic effects have varied across the globe, and there is now much discussion as to when these can have become central to the transformation of the Earth's land surface. Has this been throughout the *Holocene Epoch* (the last 11.7 thousand years), since the Industrial Revolution (in Britain for the last two hundred and fifty years, but a shorter time in many other places), or much more recently following atmospheric changes in the last several decades? Should this period now be designated the *Anthropocene* (Chapter 16)?

It is not easy to determine the timescales over which many elements of the Earth's surface have been fashioned. More than fifty years ago a fundamental concept of geomorphology (Thornbury, 1954) was that 'little of the Earth's topography is older than Tertiary and most of it no older than Pleistocene' (**Table 1.5**). Subsequent dramatic advances in dating, in geochronology (**Table 1.4**), have enabled a much greater understanding of when the Earth's surface was shaped. Many of the broad outlines are very old, including ocean basins and the core areas of continents, but many are much more recent. Different scientists analyse landscapes using a range of time resolutions from events to geological periods; many disciplines depend upon an awareness of timescales but each may concentrate upon particular aspects. It has come to be accepted that there is geological time for the geologist and archaeological time for the archaeologist, so should we have a landform time for the geomorphologist – or better still perhaps, are there simply many landform timescales that should be chosen according to their relevance for the landforms concerned? As well as being aware of society's needs, geomorphologists are driven to solve problems that they conceive to be actually solvable. This is heavily dependent on available technology – for form definition, but especially for dating.

12.1 Dividing time

Formal subdivisions for Earth history (at four hierarchical levels) are summarized in Figure 12.1, and elements from this unique programme that are relevant to landform development will be discussed more fully in Chapter 13. It is worth noting that the broad outline of these divisions emerged in the 19th century (**Table 12.1**) though revisions continue to take place. The International Commission on Stratigraphy is the body that sets this formal framework of geological phases and their boundaries. The Quaternary Period is now recognized as having lasted for the last c.2.6 Ma; the Pleistocene Epoch is divided into stages, and by convention also into Lower (c.2.6–1.8ka BP), Middle (1.8–0.8ka BP) and Upper (0.8–0.12Ka). Depositional units are generally established using international stratigraphical procedures at type sites and hierarchical name-systems involve formations, members, beds, etc. (Murphy and Salvador, 1999).

For the Quaternary, marine isotope stages (MIS) are now used to supplement and extend the classic Alpine and related regional stage equivalents (Table 12.2). The quasi-continuous ocean sediment sequence established by Emiliani (1966) was at first thought to represent temperature fluctuations, but subsequently was related to ice volume (Shackleton and Opdyke, 1973), and was thus recognized as a surrogate record of ice sheet extension and retreat. The identified 'stages' – 22 identified by Shackleton and Opdyke but now around 100 within the Pleistocene – are selectively numbered peak and trough points, numbered unconventionally from the youngest back in time. Before Stage 5 these are even-numbered for 'cold' peaks and odd for 'warm'; they are not strictly time periods with boundaries for start and end dates. A different system is used for terrestrial loess sequences, whilst lake sediments and speleothems also offer up long-term sequences, representing different aspects of environmental change rather than the ice advances and retreats that occupied only part of what were the dominating cold climates of the Pleistocene. For terrestrial sediments, cold and warm stages may be subdivided by date (chronostratigraphy), vegetation (biostratigraphy), sedimentary character (lithostratigraphy) and palaeomagnetics. Ice cores also yield diagnostic sets of trace elements and gases, radioactive isotopes, dust, pollen spores, and other characteristics that may contribute both to sequencing and dating. Unusual events such as tephra layers and episodes with ice-rafted sediments in ocean deposits similarly help to establish sequences in time. The Quaternary is an enormously rich, and often controversial, area of active research (Pillans and Gibbard, 2012;

Eon	Era	Period		Epoch	Ma		Life forms
Phanerozoic	Cenozoic	Quaternary		Holocene	0.01	Age of Mammals	Modern humans
				Pleistocene			Extinction of large mammals and birds
		Teriary	Neogene	Pliocene	2.6		Large carnivores
				Miocene	5.3		Whales and apes
			Paleogene	Oligocene	23.0		
				Ecocene	33.9		
				Palaeocene	55.8		Early primates
	Mesozoic	Cretaceous			65.5	Age of Dinosaurs	**Mass extinction** Placental mammals Early flowering plants
		Jurassic			145.5		First mammals
		Triassic			199.6		**Mass extinction** Flying reptiles First dinosaurs
	Palaeozoic	Permian			251	Age of Amphibians	**Mass extinction** Coal-forming forests diminish
		Pennsylvanian			299		Coal-forming swamps variety of insects First amphibians First reptils
		Mississippian			318.1		
		Devonian			359.2	Fishes	**Mass extinction** First forests (evergreens)
		Silurian			416		First land plants
		Ordovician			443.7	Marine Invertebrates	**Mass extinction** First primitive fish Trilobite maximum Rise of corals
		Cambrian			488.3		Early shelled organisms
					542		

Figure 12.1 The geological timescale

Elias, 2013). Correlating stages from region to region, and between the marine and terrestrial record, is often challenging, and the need for numerical dating methods (**Table 1.4**) becomes paramount.

Table 12.2 Nomenclature for the later Quaternary

Alpine	N.America	N.Europe	UK	S.America	Gl./Intergl.	Period BP (ka)	MIS
	Holocene	Holocene	(Flandrian)		I	0–12	1
Würm	Wisconsin	Weichselian	Devensian	Llanquihue	G	12–110	2–4 & 5a–d
Riss/Würm	Sangamonian	Eemian	Ipswichian	Valdiva	I	110–130	5e
Riss	Illinoian	Saalian	Wolstonian	Santa Maria	G	130–200	6
Mindel/Riss	Pre-Illinoian	Holstein	Hoxnian		I(s)	200–300/80	7,9,11?
Mindel	Pre-Illinoian	Elsterian	Anglian	Rio Llico	G(s)	300/80–455	12?
Günz/Mindel	Pre-Illinoian		Cromerian		I	455–620	13–15
Günz	Pre-Illinoian	Menapian	Beestonian	Caracol	G	620–680	16

In archaeology and history, procedures are less formal than the strati-graphical ones of geology, and names and timing vary greatly from region to region. Timescales are commonly set by documentary evidence and by ^{14}C or other dating techniques. Some 'ages' that are appropri-ate to Western Europe are given in **Table 12.3**; they may be compared with biology-based subdivisions for the Holocene in **Table 12.4**. These different phases – geological, biological and historical – have involved a range of *allogenic* or extrinsic system 'drivers' which set variable con-texts within which landforms have developed through the operation of marine, coastal, glacial, aeolian and fluvial processes. In subsequent chapters, considerable attention will be given to the later Quaternary and Holocene, and to shorter periods of anthropogenic influence when landform development has intersected with human activity.

An important additional type of temporally-set processes are *autogenic* or *intrinsic*: the trajectories and timescales in which geomorphological pro-cess systems transform the Earth's surface into recognizable form patterns, such as river valley networks, slope geometries, beach forms or glaciated topography. Within the context of precipitation/river flow and compara-ble regimes, these also have their own timescales in achieving characteris-tic forms or 'ideal' sequences under constant environmental conditions. As outlined in Chapters 6–8, they may be seen as evolving in complex ways to some defined end state (Phillips, 1999a; 2003), or possibly as achiev-ing some sort of time-extended equilibrium state or *stasis* in which forms may stay much the same, linked to on-going processes with relatively little regard for initial or eventual conditions (Gilbert, 1877; Hack, 1960). As will be discussed later (Chapters 14 and 15), several such pathways have been theoretically conceived although specification remains difficult, i.e., how to tie them down precisely, to calibrate them with dated field evidence, or even to 'validate' models for them as being truly representa-tive of real-world forms. But such concepts – temporal models and the running times for their constituent elements – are nevertheless vital to our understanding. Both physical modelling in the laboratory (Schumm et al., 1987) and computer-based numerical modelling (Anderson, 1988; Van De Wiel et al., 2011) have emerged in recent decades as being particularly useful in suggesting what may occur, models being either 'ideal' ones run without external allogenic changes, or run with changing external inputs in a controlled manner to see what the results might be.

An enormous range of timescales is involved in geomorphological research, 10 orders of magnitude in years, 17 orders defined in seconds (Brunsden, 1996). On ancient continental land surfaces, as in Western Australia, it is possible to trace river drainage systems that date to before the breakup of Gondwana in the early Jurassic about 180 million years ago, whereas the human control of many major river discharges and the transformation of their channel patterns often date to the last few decades of the 20th century. The effects of both are written into the landscape.

A river engineer may wish to know about likely change over the potential lifetime of built structures (c.100 years), but the interpretation of ancient continental landscapes in Africa and South America requires our attending to what has happened over many millions of years.

To deal with the timescales required by the geomorphologist three separate 'times' were distinguished by Schumm and Lichty (1965):

- Cyclic or geologic time – encompassing millions of years, for example to complete an erosion cycle.

- Graded time – which may be hundreds of years during which a graded condition or dynamic equilibrium exists.

- Steady state time – typically of the order of a year or less when a true steady state situation may exist.

This paper – reviewed as a classic (Kennedy, 1997: 420) not only for the recognition of these three timescales, but also because the status of geomorphic variables depended on the timescale being investigated – has been identified, together with the magnitude/frequency concept (Chapter 10), as providing the main conceptual basis of thinking up until 1985 (Tinkler, 1985). This approach was vital in attempting to reconcile and relate timeless and time bound (Chorley and Kennedy, 1971: 251) approaches to geomorphology.

Schumm and Lichty (1965) linked timespans to the dependence or independence of 10 variables like initial relief, vegetation or morphology. The concept was put forward as statistical procedures were coming to the fore in geomorphological research, and when states of equilibrium (Chapter 6) were deemed to reflect an operational balance between some variables, with others being irrelevant at particular study timescales. Quantitative relationships could be established through correlation and regression procedures. The identified timescales were neither calibrated nor absolute – for example, with badland topography or drainage networks on an intertidal beach developing over shorter timespans than fluvial relief forms on resistant rock. But the scene was set for establishing the relevance of measurable factors at different timescales, and for reconciling long-term evolutionary geomorphology with studies that recognized short-term equilibrium states.

Earlier geomorphologists had little access to effective dating tools, and in fact envisaged morphology itself as bridging a kind of evidence gap between the observable present and the geological record. Thus W.M. Davis (1899) proposed evolving relative stages that could be read from surface forms to set a temporal framework for landform development (youth, maturity and old age; see Chapter 11) but the timescales were largely unknown. Many techniques for dating both deposits and exposed rock surfaces have subsequently become available (**Table 1.4**) so that this perceived evidence gap is now less apparent.

12.2 Events and episodes

So far in this chapter, temporal frameworks have been presented in two forms – either on a continuous timescale in years, or as discrete episodes in Earth history. Historical studies of most kinds commonly do this: we use terms like 'the Middle Ages' and also give dates in years (AD/BC or CE/ BCE). It is worth considering these alternatives a little more closely. Years, days and lunar months have physical underpinnings (based on the circulation of the Earth around the sun, the Earth's rotation, and the phases of the Moon), whereas hours and minutes are arbitrary. It was the Babylonians (c. 300 BC) who divided the day sexagesimally (that is into 1/60ths). 'Hour' comes from a general word for time in Greek via Latin; the 12 hours of the day for monks in medieval times were not fixed in length but varied with the season. Nevertheless these briefer time periods have pragmatic human meaning. Once they are fixed they are equally useful in a scientific context, for example in expressing river velocity in m s^{-1}. All base units and units derived from them are now strictly defined in an International System (SI), as in units of length and mass, and derived units such as force, power and energy (Chapter 9). The *second* is the base unit of time, as proposed by Gauss in 1832, but now determined using an atomic clock.

But usefully envisaged episodes, like the *Little Ice Age* (LIA, c.1350 – 1850 AD) (Grove, 1988), are less easy to tie down: these are conceptual entities that may contain internal variety, and they have start and end dates that are matters of study and opinion. The LIA climate was driven at different times by minima in solar energy output, by atmospheric circulation variations, and by the effects of volcanic eruptions on atmospheric composition. Was it really a single 'Age'? The inhabitants of alpine areas, threatened by crop failure, rockfalls and glacier advance, may well have thought they were living through distinctively unfortunate times, though conceived as a historical period this was only recognized in retrospect. In archaeology, similarly designated eras like the *Palaeolithic* or the *Bronze Age* (**Table 12.3**) are in general use, though they may be *time-transgressive* or *diachronous* in that they start or end at different times in different parts of the world. Other happenings may be more in the nature of singular events or climaxes potentially susceptible to precise dating, such as the *Last Glacial Maximum* (LGM) taken in northwest Europe to be around 20,000 years ago when ice sheets most recently reached their greatest extent. Sea levels were lowered, permafrost was extended, and global vegetation cover was transformed. But this 'event' was not coincidental everywhere, nor in practice is the LGM as yet that well defined globally. It appears that the timing of maximum advance varied very considerably worldwide (Hughes et al., 2013) so that a term like the LGM is not a chronologically meaningful one.

Other 'event' examples include catastrophic discharges arising from the sudden refilling of the Mediterranean basin after its desiccation in the Messinian (in the Miocene 5.9 to 5.3 million years ago); the initial breaching of the Straits of Dover from a glacially ponded lake in the southern North Sea (probably 400,000 years ago); the Dansgaard-Oeschger and Bond events discussed in Chapter 8; and the sequence of great Missoula floods in western America arising from drainages from another ice-dammed lake at the end of the last glaciation around 15,000 years ago. These are extreme examples, and the definition of an 'event' may be scale dependent, to include a whole glacial-interglacial cycle or merely an occurrence such as a rainstorm or boulder impact (see Brunsden, 1996, for an extended discussion of the range and nature of events in geomorphology).

On the short timescales relevant to observations of many physical processes, such parameters as temperatures, precipitation or river discharges will vary daily, hourly or even second by second in terms of turbulence patterns. They may be expressed in terms of continuous time series covering periods for which records have been extracted. It may be preferable for some purposes to use average values such as mean daily, monthly, or annual temperatures or river flows. Attention in geomorphology has also focused on frequency/magnitude concepts (Chapter 10). Given a reasonably long run of data, extreme events assume particular statistical distributions and it is possible to estimate their likely probability, as in the recurrence interval, in years, between large floods. Thus the flood magnitude with a probable annual recurrence interval of $1/100$ ($P = 0.01$ or a 1% chance) is popularly thought of as the 'one hundred year flood'. This emphatically does not mean that there will be a one hundred year time gap between two equivalent events. It also assumes technically that the data forming the basis for analysis are all from a single statistical population; if conditions have changed (for example, when climate drivers produce clustering of events, or more extreme conditions) then frequency estimates based on misplaced statistical distribution assumptions lose their meaning. Nevertheless, in geomorphology considerable attention has been given to events (in magnitude and frequency terms) that produce the most 'work'. These may be regarded as *formative events*, with the perhaps misleading implication that others are not. The size of river channels has been related to a 'bank-full' river discharge with a frequency of between one and two years (Chapter 11), though with some variation (Williams, 1978). This does not work so well in some environments, and there are limitations arising from the use of only short runs of data and the non-inclusion of rare events that may do work that is different in kind, such as causing major landslides (Lewin, 1989). What may be regarded as 'regime' conditions – the full range of operational processes – normally include seasonal variations and events for such aspects as rainfall and ice accumulation. It is also common practice to measure geomorphological processes directly, but necessarily over a short time period, and then link these to long-term development by calibrating them against other parameters that have been recorded, such as climate or river discharge data.

Both individual events/episodes – including the sequence of glacial and interglacial episodes of the Pleistocene – and their frequencies may serve to highlight a challenge that geomorphology has to face when dealing with the development of landforms (Figure 12.2). Allogenic (extrinsic) drivers may not remain constant for long enough for autogenic (intrinsic) processes to achieve characteristic forms, and this is especially true of longer-term landform development where millions of years may be involved for landform relief reduction. Figure 12.2 shows fluctuations in climate over the last 540 million years of Earth's history: there have been several ice ages but also warm periods, as in the Cretaceous. Evidence for the longer-term history of landforms can sometimes amount to a surrogate catalogue of Earth system change alone. Such evidence may be recorded as fragmentary and fortuitously preserved sedimentary bodies left isolated after the

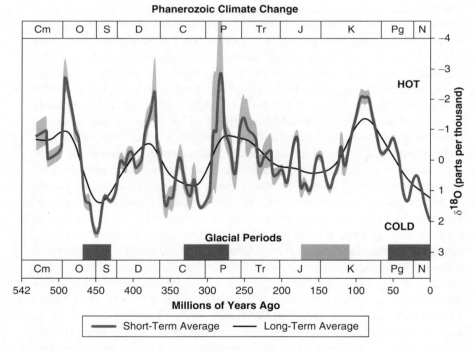

Figure 12.2 Climate change over the last 540 million years

Both formal geological phases and a continuous timescale are given. The solid black line gives a smoothed climate trend. Note the glacial periods prior to the Pleistocene in the Neogene (N), and the warm periods as in the Ordovician (O), Permian (P) and Cretaceous (K). There was a general cooling thereafter. These subsequent changes and episodes are considered in Chapter 13 (see Figures 13.2–13.4).

episodes to which they bear witness. How any largely erosional landscape developed during or in between such times by autogenic processes may be much more difficult to establish. By contrast, physical processes measured over a few years (as in much postgraduate research, and in short-term customer-funded contracts) may be vital in sorting out dynamic processes and their drivers, but they may or may not be representative of a longer process suite which incorporates unobserved but formative extremes (such as glacial surges or large floods) or the changing trajectories that arise from extrinsic allogenic change.

12.3 Conclusion

For understanding landforms it is necessary to refer to a range of time-scales or classifications of time such as formal geological epochs, conventionally accepted Quaternary chronozones, or terms like *The Little Ice Age*. Geomorphologists have also invented classifications of time such as the steady, graded and cyclic time suggested by Schumm and Lichty (1965). Such distinctions are required because research on landforms necessarily refers to time periods of different duration: process investigations may focus on a few years (the 'steady time' of Schumm and Lichty) whereas remnants of ancient land surfaces that can be identified by cosmogenic dating required millions of years (the 'cyclic time' of Schumm and Lichty). Timescale terms used in landform investigations now include instant/now/instantaneous time (a few seconds up to years); steady or event time (up to many years); management/engineering time (years to centuries); 'graded' time (thousands of years); and cyclic/geologic/evolutionary/phylogenetic time (millions of years). These are all inventions to assist the interpretation of landform origin, and the continuing challenge is to relate research results obtained from one such timescale to investigations conducted at another timescale. It is now both possible and necessary also to use a variety of dating techniques (**Table 1.4**). In earlier studies of landforms this was not feasible so that studies of active processes were conducted independent from studies of landform evolution (Chapters 6, 8 and 11). However, recent developments, particularly in modelling and through cosmogenic and other dating methods, have enabled improved cross-temporal investigations to be undertaken, and actual as opposed to theoretical timescales to be used – one of the most exciting developments in the recent investigation of landforms.

FURTHER READING

Bell, M. and Walker, M.J.C. (2005) *Late Quaternary Environmental Change*. Harlow: Pearson/Prentice Hall.

Bishop, P. (2007) Long-term landscape evolution: linking tectonics and surface processes, *Earth Surface Processes and Landforms,* 32: 329–65.

Brunsden, D. (1996) Geomorphological events and landform change, *Zeitschrift für Geomorphologie,* NF 40: 273–88.

Pillans, B. and Gibbard, P. (2012) The Quaternary Period. In F.M. Gradstein, J.G. Ogg, M. Schmitz and G. Ogg (eds), *The Geological Timescale.* Amsterdam: Elsevier.

Schumm, S.A. and Lichty, R.W. (1965) Time, space and causality in geomorphology, *American Journal of Science,* 263: 110–19.

Woodward, J. (2014) *The Ice Age: A Very Short Introduction.* Oxford: Oxford University Press.

TOPICS

1. Construct a table of landforms, such as moraines or coastal cliffs, allocated according to the different time ranges for which they would need to be studied to establish their development.

2. What parts of the Earth's surface might or might not be expected to provide good evidence for stages in long-term landform autogenic evolution, independent of external environmental change?

 ## WEBSITE

For this chapter the accompanying website **study.sagepub.com/greg oryandlewin** includes Tables 12.1, 12.3, 12.4; and useful articles in *Progress in Physical Geography.* References for this chapter are included in the reference list on the website.

13

FORCINGS

Forces external to systems that result in system responses are known as forcing functions or drivers. These are responsible for environmental change that can occur over a range of timescales. Relief sets available energy, and tectonic mobility can alter systems whilst exogenetic processes are in operation. Surface forces can be natural or human-induced and can cause change directly or indirectly. Whereas a direct forcing function or driver unequivocally influences system processes, an indirect driver will operate more diffusely by altering one or more direct drivers.

From a geomorphological perspective, dated environmental drivers and changes include: tectonic activity; glacierisation, periglacial and other morphoclimatic systems beyond glacial limits; vegetation changes; eustatic sea level changes; and isostatic crustal adjustments. Characteristically these exhibit lagged relationships with landform responses, and there are transitional states as conditions change. For example, ice sheet withdrawal following climate amelioration may involve little vegetation ground cover and enhanced erosion until plant colonization has had time to take place. Today there remains something of a conceptual dilemma: are there some episodes that require identity as particularly significant geomorphologically, or is it best to focus on the nature of multiple response trajectories? Both have validity, but these alternative approaches also incorporate beguiling conceptual implications.

13.1 Allogenic histories

Theoretical models of long-term landform evolution used to assume 'instantaneous' uplift as the starting point for the development of landforms (Chapter 11); however, advances in tectonic understanding show that crustal mobility is continuous, if generally slow, but with uplift very unevenly distributed across the Earth's surface. Thus landform development can be coincidental with uplift, with the removal of mass from uplifted areas, sometimes contributing to further isostatic uplift. This is an enormous area of geophysical investigation (Grotzinger et al., 2006; Allen and Allen, 2013), with implications for geomorphological modelling. **Figure 13.1** shows dynamic changes in elevation over 6–45 million years across South America, driven by mantle convection just as the major landforms of the Andes were developing through both tectonic and erosional activity.

For the Middle and Late Quaternary, Figure 13.2 summarizes the set of glacial and interglacial marine isotope fluctuations and stages that were established in the later 20th century. A conceptual revolution took place when it became apparent that there were many more environmental fluctuations than previously revealed by the fragmentary land-based evidence. There has been a history of Cenozoic cooling since the Eocene, with the onset of significant Northern Hemisphere glaciation at about 2.6 Ma (c. MIS 100) near what is now taken as the start of the Quaternary (Figure 13.2) (Ehlers and Gibbard, 2008). At c.1.2–0.8 Ma (or MIS 22) there occurred the 'mid-Pleistocene transition', with the domination of larger-scale c.100 Ka climatic cyclicity, a set of major glacial episodes, and large changes to extraglacial erosional environments involving deep valley incision (Gibbard and Lewin, 2009). For most of the last million years conditions have been colder than present, with interglacials (like the present Holocene) being shorter in length and not the 'norm' for landform development (Porter, 1989). Both the cold and the cyclic rhythm have imposed a distinctive forcing character on Quaternary geomorphology (Bridgland and Westaway, 2008), though there is much more to resolve as to what actually happened when.

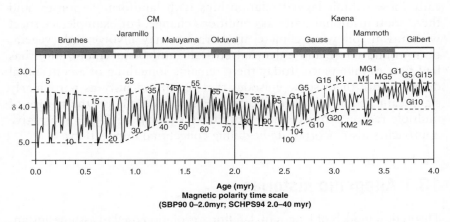

Figure 13.2 The marine oxygen isotope signal (MIS) for the last 4 million years

Age in million years (bottom) and magnetic polarity (top), showing marine oxygen isotope signal. MIS stages are numbered back to c.2.6 Ma; earlier fluctuations are labelled according to their magnetic polarity (after Shackleton, 1995).

In detail, the picture becomes even more complex (**Figure 13.3**). Within the most recent glacial-interglacial cycle, there have been multiple 'stadial' peaks as seen in marine sediments, with some identified more broadly in terrestrial deposits (Van Huissteden et al., 2001). For much of the time there were cold conditions but not necessarily ones of glacial extension, whilst some 'cold' stages had several glacial advances and deposits. Figure 13.3 shows that cold episodes also exhibit a long saw-tooth or zigzag decline but with abrupt 'terminations' which may take less than 50 years. Such transitions have attracted particular geomorphological attention (Vandenberghe, 2008). Ice sheet advances and recessions during a glacial-interglacial cycle have also proved to be complex, responding both to climatic fluctuations and internal glacier dynamics. There may be both fast flowing and less rapidly moving ice streams spreading from separate ice centres during a glacial cycle. Surges and advance/retreat cycles on centennial timescales correspond to climatic forcing that may produce alternate relatively cold-based and wet-based ice episodes (Hubbard et al., 2009).

Although they may not show the same degree of change, interglacials like the Holocene can also be subdivided into stages: biologically via pollen zones (though these may be time transgressive) (**Table 12.4**); according to eustatic sea level rise and isostatic recovery (variable with former ice cover extents); or in geomorphological process terms (Gibbard and Lewin, 2002). Analysis of European river sedimentation episodes suggests complex regional and climate-proxy matching patterns (Macklin et al., 2006).

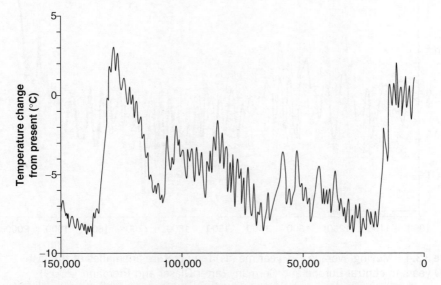

Figure 13.3 The last cold stage, and its intricate trace of climatic change (from Petit et al., 1999)

Finally, the last thousand years – when historical and instrumental climatic data have become available (Figure 13.4) – show the effects of anthropogenic impacts on vegetation and erosion more clearly (**Figure 13.5**). Post-Enlightenment human impacts can also be documented (Figure 13.6), so that sub-centennial environmental changes become very apparent (Glaser and Riemann, 2009; Lewin, 2013). What could be called geomorphologically effective 'disturbance regimes', tracking the changes of human agency, may be complex, involving both direct effects such as the construction of embankments and ones mediated through changes in vegetation or cultivation which lead on to changes in sediment dynamics. These may be 'indirect drivers' that, by changing others, lead on to geomorphological impacts (Nelson et al., 2006). Reinhardt et al. (2010) have emphasized the need for geomorphologists and ecologists to work together as their subject matters are co-evolutionary. Biological effects can also be stabilizing or destabilizing (Viles et al., 2008).

Recent centuries have seen many interacting changes; for land-forming processes these may be critical, whilst on-going variability sets the scene for the climate change concerns of the last few decades. Constant 'regime' conditions, involving fluctuations around a mean, may be assumed for design purposes and for short periods (with an allowance for variability and extremes).

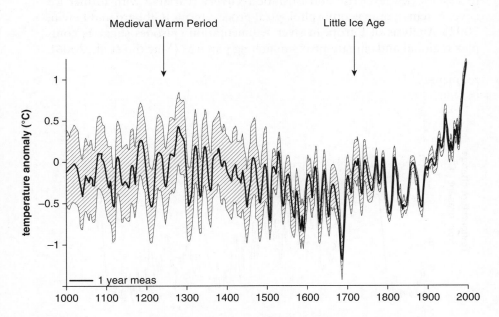

Figure 13.4 Moving-average (11 year means) temperature anomalies of the last 1000 years in central Europe and Germany (after Glaser and Riemann, 2009)

The plot uses historical data (AD 1000–1750) and instrumental data (AD 1751–2007), and the shaded area represents a measure of likely error.

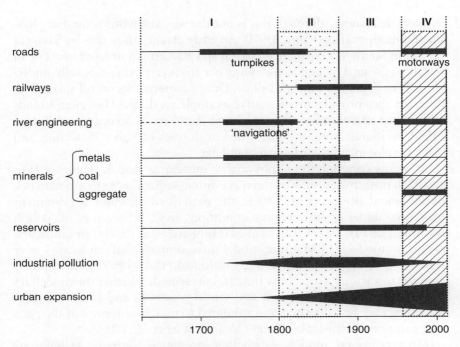

Figure 13.6   Post-Enlightenment fluvial modification drivers in the UK. I–IV show major periods of transformation. (after Lewin, 2013)

But the now-known evidence suggests that variability is a much greater environmental characteristic than was appreciated just a few decades ago. For the longer term, geomorphological attention may focus on particular forcing episodes, whilst the identification of transitional states like *paraglacial* ones – those set by and following prior glaciation – appear particularly important in many mid-latitude temperate environments (Ballantyne, 2002a). For the Holocene and present-day environmental management, intermittent more extreme episodes and recovery from them, together with the complex interventions of human activity, have been the focus of much attention.

13.2 Autogenic processes

How may autogenic time trajectories be determined? For shorter-term land forming activity this may be via direct observation, and then for somewhat longer by the use of historical documentation – with maps and remote sensing information being particularly useful. But the longer the timescale, the greater the fragmentary nature of field evidence. One solution is to find situations in which a set of co-existing forms can be assumed to represent a time sequence. This is known as 'space-time substitution' involving *'ergodic'* reasoning, although this is not the way the word is used in physics and mathematics (Paine, 1985). An early classic study was by Savigear (1952) in South Wales where a coastal spit had grown in front of a line of sea cliffs, giving a 'young to old' sequence from vertical and basally-undercut cliff to degraded subaerial slope. Drainage networks on till surfaces of different ages provide an alternative example, as do the Hawaiian Islands where a set of volcanoes developed and then became extinct in sequence, island by island. These exhibit different degrees of valley dissection and slope development according to island age.

It is now common to use physical or numerical models (Figure 9.1) to establish trajectories, such as the space-filling sequence of stream network development (Rigon et al., 1993), the pattern of meander development over time under specific regime conditions (e.g., Schumm et al., 1987; Nicholas, 2013), ice sheet dynamics (Hubbard et al., 2009) or landscape evolution models (LEMs) that model three-dimensional landscapes over time (e.g., Willgoose et al., 1991; Coulthard et al., 1999). Such models give rise to various issues: how time is represented; whether the modelling processes adequately represent real-world processes and situations; and how they may be calibrated and validated to match or represent the pace and nature of Earth-scale change (Van De Wiel et al., 2011).

Most temporal models – whether presented verbally, visually or numerically – involve development tracking from an initial to an end state. This may be one of constant form as material continues to pass through the system, steady state equilibrium, or dynamic equilibrium if the equilibrium state itself has an inbuilt trajectory (Chapter 6). This can be viewed as a form of self-organization (Phillips, 1995b). Alternatively things may move to a terminal 'decay' state, when land-forming processes are exhausted and the energy to accomplish work is no longer available (as in static equilibrium under conditions of maximum entropy).

Both actual and modelling timespans may vary considerably; in fact one of the factors behind adopting a modelling strategy is the impossibility of waiting long enough for landforms to change sufficiently in the real world to allow observations of how they will actually develop. Thus it is possible to observe 'river' channel development in a sandpit or to use modelling as a kind of controlled experiment under a few critical operational conditions rather than what may seem to be the anarchic mixture of field processes. Such 'reduced complexity' models may nevertheless achieve realistic representation of field developments – insofar as these can be validated (Chapter 14).

13.3 Thresholds and perturbations

The threshold concept is common to many sciences: thresholds are points marking a transition from one state to another, one such point being that at which change or movement starts to take place in response to driving forces. Thus in physics an applied force has to overcome resistance before movement occurs so that there is not a simple linear relationship between force applied and movement response. In sediment transport, the force applied by wind or water – often expressed in terms of velocity, shear stress or power (Chapter 10) – must overcome others (such as friction or gravity) which tend to oppose movement, and there is a threshold that must be exceeded before this occurs. With a decrease in applied force, resistance forces may lead to the cessation of movement – deposition rather than the 'entrainment' that occurred at the transport-onset threshold in the case of sediment being moved by flowing water. These may not be the same, for example because sediment of a given size may be packed into a channel bed requiring a larger force to get it moving.

Particularly following the work of S.A. Schumm (1973; 1979), this concept has been applied to **geomorphic thresholds**. Schumm divides thresholds into 'extrinsic' and intrinsic' ones generated in relation to allogenic and autogenic drivers, with the latter perhaps being 'developed within the geomorphic system by changes in the morphology of the landform itself through time' (Schumm, 1979: 487). He exemplifies these using the development of valley-floor gullies, once critical threshold gradients are exceeded, and by stream meandering in physical model studies observed to begin above threshold slopes and stream powers. Thresholds can be ones involving process, perhaps from sediment stability to motion, but also produced by morphological change itself. These Schumm envisaged as a separate category of 'geomorphic' thresholds (Chapter 7). Schumm (1969) also called 'an almost complete transformation of river morphology' a **metamorphosis**. More generally this may be described as a state transition: 'a change that results in a qualitatively different landform, geomorphic environment or landscape unit' (Phillips, 2014).

Developed at the same time was an important general concept (Graf, 1977) that used an analogy with the radioactive decay rate law to model the change from one equilibrium situation to another. This rate law concept, which could embrace several types of equilibrium, proposes that – after a disruption such as a change in climate – a reaction time begins when the system absorbs the impact, but subsequently during the relaxation time there is adjustment of the system towards a new equilibrium condition. This approach, consistent with temporal sequences proposed by Knox (1972) and more recent approaches to disturbance regimes (Viles et al., 2008), is elaborated in Chapter 14.

Extrinsically-driven geomorphological change might seem easiest to distinguish where there has been a threshold-crossing change of a morphogenetic system on a very large scale, for example as between fluvial or glacial regimes. Earlier geomorphological efforts were often concerned with such overarching extreme changes: were temperate landscapes indeed moulded by prior glaciation, not once but several times; were ancient planation surfaces terrestrial or marine in origin; or were present-day landscapes in mid-latitudes largely relict from prior periglacial or even tropical process environments? The accumulation of environment-diagnostic criteria for former morphogenetic systems in the sediment and weathering residue record has furthered this process. Inheritance (Chapter 15) is impor- tant because allogenic changes overlap the evolutionary timescale for 'complete' autogenic transformation, and multiple relict forms (like planation surfaces, terraces and diagnostic materials from earlier sys- tems) persist in at least parts of many landscapes (Ebert, 2009).

Over short timescales (years to millennia) it is more difficult to distin- guish within-system, but extrinsically forced, changes from intrinsic ones since they are worked through by modifying rather than transforming regimes that are, in any case, in a state of flux. Again, how are extrinsic and intrinsic to be separately defined, for example in the case of an extreme flood? Increasing the frequency and magnitudes of floods beyond those regarded as being an integral part of contemporary regimes may enlarge river channel size or the dimensions of meanders, thus providing evidence of environmental change (Knox, 1993). But modelling studies suggest that whilst an external forcing event may trigger on-site erosion, the on-going transfer of material through cascading systems as a whole may be differ- entially time-consuming and complex so that it can blur the down-system effects (Van De Wiel and Coulthard, 2010). In river catchments, variable slopes, network travel distances, storage involvement and connectivity all contribute to this potential eventuality. In human-modified landscapes, it depends what, where and when these modifications were in terms of their potential to transform the larger landscape. The same applies to ice sheet dynamics and their response to climatic forcing. Responses may also involve complex interactions, as when a decrease in precipitation may reduce vegetation cover but enhance dune mobility and slope erosion. A number of conceptual response models of this kind in biogeomorphologi- cal systems have been reviewed by Viles et al. (2008).

Outcomes also depend on the nature of the driver: is it a brief event or *perturbation*, or a more permanent change that lasts long enough to have effects that permeate throughout the system in question to become a new 'norm'? Brief recent events whose impacts have been studied include rare extreme floods, drought episodes, catastrophic lake discharges, forest fires, volcanic eruptions, tsunamis, and mining activity (Chapter 17). Many may be regarded as natural hazards (see **Table 17.2**). On a longer timescale, cli- matic episodes at around 8.2Ka and 4.2Ka have been suggested as markers for dividing the Holocene into Early, Middle and Late (Walker et al., 2012). Other researchers have suggested anthropogenic transformations that, at least as far as we can see, have permanently changed morphogenetic systems (Chapter 16). Bunge (1973) has made the analogy that 'cities are like karst topography with sewers performing precisely the function of limestone caves in Yugoslavia, which causes a parched physical environment, especially in city centres'. Changes such as this or the effects of dams on sediment systems may be rapid but are not expected to be reversible (Downs and Gregory, 2004). Perturbations may cause deviation from what may be presumed, rightly or wrongly, to be their former equilibrium or regime state (Chapter 6), but then reversion. Prior states may be established by a pre-event survey or behaviour over a reference period. As previously discussed, perturbation studies may usefully involve tracking the response, relaxation path and timing for recovery from such events or episodes, which are often extrinsic in origin but have also included extreme events (like very rare extreme floods) within the range of variability of existing systems.

13.4 Conclusion

Some of the difficulties of distinguishing relevant drivers discussed in this chapter arise from the large range of spatial and temporal timescales involved in geomorphological research. Schumm and others (Schumm and Lichty, 1965; Church and Mark, 1980) have shown that variables may be regarded as independent drivers at one timescale but become intrinsic and dependent at another. Tectonics may be regarded for study purposes as irrelevant in event studies of rill erosion, but drivers of long-term landform development or landslide instability in other cases. Tectonics may not *actually* be unimportant as it determines relief energy but does not need to be considered for many short-term process studies. Nevertheless there is a real history to environmental drivers from parallel Earth systems, including anthropogenic ones, and their changes may be superimposed on the geomorphological process-response systems that would otherwise take place under regime conditions. Reversing the picture, the effects of allogenic drivers are also mediated through intrinsic processes, and in geomorphological systems the existence of autogenic processes means that changes are not simple responses to external drivers such as climatic change or anthropogenic activity.

FURTHER READING

Bridgland, D. and Westaway, R. (2008) Climatically controlled river terrace staircases: a worldwide Quaternary phenomenon, *Geomorphology*, 98: 285–315. DOI: 10.1016/j.geomorp.2006.12.032

Ehlers, J. and Gibbard, P. (2008) Extent and chronology of Quaternary glaciation, *Episodes*, 31: 211–18.

Gregory, K.J. and Goudie, A.S. (2011) *The SAGE Handbook of Geomorphology*. London: Sage (see especially Chapters 3 and 4).

Lewin, J. (2013) Enlightenment and the GM floodplain, *Earth Surface Processes and Landforms*, 38: 17–29. DOI: 10.1002/esp.3230

Macklin, M.G., Lewin, J. and Woodward, J.C. (2012) The fluvial record of climate change, *Philosophical Transactions of the Royal Society,* A 370: 2143–72.

Phillips, J.D. (2009) Changes, perturbations, and responses in geomorphic systems, *Progress in Physical Geography,* 33: 17–30.

Schumm, S.A. (1979) Geomorphic thresholds: the concept and its applications, *Transactions of the Institute of British Geographers,* NS 4: 485–515.

Schumm, S.A. and Lichty, R.W. (1965) Time, space and causality in geomorphology, *American Journal of Science,* 263: 110–19.

Viles, H.A., Naylor, L.A., Carter, N.E.A. and Chaput, D. (2008) Biogeomorphological disturbance regimes: progress in linking ecological and geomorphological systems, *Earth Surface Processes and Landforms,* 33: 1419–35.

Walker, M. and Lowe, J. (2007) Quaternary science 2007: a 50-year retrospective, *Journal of the Geological Society of London,* 164: 1037–92.

TOPICS

1. How do you think the Holocene should be subdivided, taking into account the changing set of drivers that have influenced landform development over this period?

2. List the direct and the indirect drivers that might lead to a change in glacier extent or stream channel sizes and patterns.

 WEBSITE

For this chapter the accompanying website **study.sagepub.com/ gregoryandlewin** includes Figures 13.1, 13.5; and useful articles in *Progress in Physical Geography*. References for this chapter are included in the reference list on the website.

14

CHANGE TRAJECTORIES

There have been many attempts to track the change sequences that result from form development, including responses to climatic and other disturbances. These may be conceptualized diagrammatically or they may derive from physical or mathematical modelling. Some research suggests that change sequences have multiple paths and outcomes; others propose that the end results from different processes are very similar. Complexity characterizes many systems, partly because of their hierarchical nature and also because the allogenic contexts or spatial configurations in which active systems operate are very varied.

Previous chapters (6, 7, 8 and 11) have shown how landforms have been conceived historically as being in 'timeless' equilibrium states, as subject to directional or cyclical change, or as being complex in terms of both their assemblages and the routeways of change they follow. The nature of recent Earth history has also imposed a series of disturbances on such systems, both because of 'natural' climatic and allied fluctuations, and because of growing anthropogenic effects (Chapters 13 and 16). This chapter focuses on how the understanding of such disturbances has been approached and modelled (14.1), how multiple possible outcomes complicate predictions, and how several paths may lead to similar outcomes (14.2). The hierarchical nature of geomorphological systems also introduces complexities (14.3). All modelling necessarily involves a range of conceptual issues (14.4) including error, sensitivity, representativeness, validation, reversibility and repeatability.

14.1 Change models

Models now available fall into several categories: these include initiation or colonization models; progressive change models; and stasis models (either dynamic equilibrium models or ones where change does not appear to take place). A range of models is presented in **Table 14.1** and may be applied to whole systems, or to the behaviour of sub-systems. They may be in qualitative form or numerically driven in computer modelling. Concepts may be presented verbally, and often in the form of a set of 'key stages' taken as landforms are transformed, but also visually as diagrams or sets of cartoons (Figures 14.1 and **14.2**). Video clips usefully allow the visualization of model runs, whether physical or numerical. However, in many areas, there have been comparatively few prediction-capable advances beyond suggesting the *direction* of response, with measured elements increasing or decreasing, as in the set of ideas presented over forty years ago by S.A. Schumm (1969) concerning channel dimensional change in response to discharge or sediment load changes (Table 14.2) which built on the proposal of Lane (1955). Quantitative specification may be possible where there are established linear relationships between control and response variables, for example meander dimensions and river

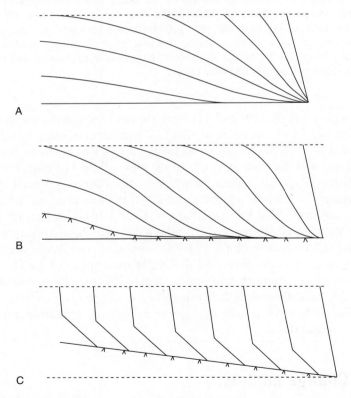

Figure 14.1   Models of slope evolution, showing equal-age stage lines, and evolving basal surfaces that are of composite age (arrowed lines)

(A) represents Davisian slope decline, (B) slope replacement as attributed to W. Penck, and (C) parallel retreat across a pediment surface as associated with L.C. King.

Table 14.2 The response of river channel variables slope (s), sediment size (D), depth (d) and width (w) to changes in the discharge of sediment (Q_s) and water (Q_w) (modified from Schumm, 1969, and Odoni and Lane, 2011: + indicates increase, ++ strong increase, - a decrease,– a strong increase, . negligible change, * an unpredictable response)

Driver	Adjustment	Example of Change
Qs., Qw++	s–, D+, d+, w++	Long-term urbanization, increased frequency and magnitude of runoff, generally leading to channel erosion
Qs., Qw+	s–, D+, d+, w+	Initial response to afforestation; general increase in discharge, widening of channel but no increase in sediment delivery (small Qw+, w+)
Qs–, Qw–	s–, D+, d+, w*	Longer-term vegetation effect, with water balance change and reduction in sediment delivery
Qs., Qw+	s–, D+, d+, w+	Increase in runoff with more extreme events: if sediment supply-limited, Qw dominates with both d+ and w+
Qs+, Qw+	s–, D*, d–, w*	Increase in runoff with more extreme events, with increase in overland flow and increased sediment delivery: system is transport-limited and Qs dominant over Qw; feedback effects on D and s difficult to predict

discharge (Knox, 1993). Another potentially predictable outcome occurs where thresholds have been established and which may be crossed, though this may not be clear-cut.

For river channel patterns, following Lane (1957) and Leopold and Wolman (1957), such thresholds may be revealed on plots of discharge versus a measure of stream power in which a threshold line may discriminate between the domains of different channel styles – particularly meandering and braiding (**Figure 14.3**). Surrogate morphological variables are catchment size and slope, and these have been substituted when measures for discharge and stream power are unavailable. On-going research has complicated the picture: channel pattern thresholds may be more in the nature of transitional zones (Eaton et al., 2010; Kleinhans and van den Berg, 2011) reflecting other factors including bank resistance and vegetation not included in a two variable analysis. Some channel patterns such as straight (or near straight) or anabranching are not easy to allocate to particular domains on bivariate plots (Lewin and Brewer, 2001; Latrubesse, 2008). In sediment transport, bed armouring (the protection of finer bed material by coarser particles and packing) may mean that theoretical transport thresholds for a given particle size have to be considerably exceeded for movement to take place, and mixed grain-size bed material becomes mobile all at once in more extreme events. Here entrainment thresholds are different from what Church (1978) called 'underloose' material.

Form sequences in time (Figure 14.1) – a kind of spatial history rather than a simple two-variable or multivariate graphed relationship – may be successfully derived from physical or numerical modelling. Models may be used to specify initiation patterns, as is illustrated by studies of river patterns, a research area in which formal modelling has been very active. Channel models follow the deformation of straight channels into braided or meandering ones, while network models track the colonization of initial land surfaces by extending and branching channels and valleys. A variety of scales may be simulated, from erosional rills developing across cultivated fields, to centuries of river channel and millennia of mountain landform evolution. They may be physical and based on flume or sand tray studies, as in the 'rainfall erosion facility' maintained at Colorado State University for several years by S.A. Schumm, his students and co-workers (Schumm et al., 1987). These echo the many engineering approaches to modelling, often being especially concerned with the design of stable rivers and irrigation canals that do not become liable to serious erosion or sedimentation. Some models involve formal hydraulic scaling so as to maintain as closely as possible a hydraulic similarity with the larger-scale counterparts they are intended to model, while others simply monitor transformations effected by sprayed and flowing water on fine-grained material moulded into channels, fans, and flat or sloping surfaces. Good examples come from the whole series of scaled experiments at the University of Utrecht by M.G. Kleinhans and his students, involving investigations of river meandering and other channel planform developments (e.g., van Dijk et al., 2012; van de Lageweg et al., 2013), whilst Nicholas et al. (2009) couple physical and numerical modelling for alluvial fans.

Numerical models have come to the fore in recent decades, particularly given the availability of large volume, high-speed computational facilities, and include models of river channel change, channel network growth and large-scale landscape development, as in the case of simulated fault-mountain evolution reviewed in Van De Wiel et al. (2011). Good examples include the pioneering study by Howard (1992) modelling meander development, and braided rivers modelling by Murray and Paola (1994) and by Nicholas (2013b) who modelled a continuum of river channel patterns (meandering, braided and anabranching) as developed for periods of >350 yr from initially straight outlines, taking into account variable grain sizes, bank erodibility and vegetation (**Figure 14.4**). Bishop (2007) and Tucker and Hancock (2010) have critically reviewed large-scale landscape evolution models (**Table 14.3**). As Tucker and Hancock observe, these models sharpen thinking by testing theory as expressed in quantitative form, they expose the logical consequences of theory application, and they stimulate new ideas. Computational sophistication is increasing all the time; for example, the development of cellular automaton models (initially developed by S. Ulam and J. von Neumann in the 1940s) has stimulated progress in understanding river and catchment development over time (e.g., Murray and Paola, 1994; Coulthard et al., 2002). A cellular automaton is a grid of cells each with a finite number of states; rules then determine new generations in terms of the current state of each cell and the states of the cells next to them. This 'neighbour to neighbour' model may or may not be applied usefully across the range of geomorphological processes.

Quantitative field evidence-based response patterns are disappointingly scarce, but some do support the general trajectories discussed initially in Chapter 6 (Figure 6.1). Following a disturbance in input-output systems there may be a lag (the *reaction time*), then a response possibly followed by *recovery*, or the achievement of a new adjustment state if the change is permanent (the *relaxation time*). Recovery may be delayed while responses may be transient, particularly where there is a succession of events without complete recovery between them (Figure 14.5).

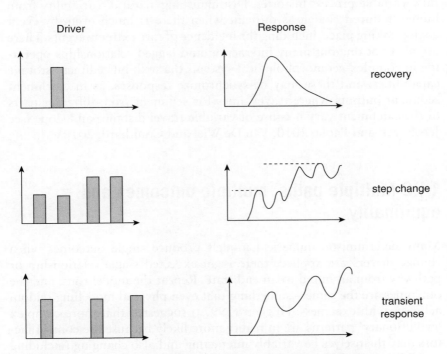

Figure 14.5 (a) Different types of driver eliciting differential responses, (a) a rapid recovery, (b) a step change, and (c) a transient response

There may be one major flood affecting channel dimensions or changing channel pattern entirely, or there could be a succession of such events (Burkham, 1972). **Figure 14.6** shows river channel responses to sets of flood events subsequently 'frozen' because of upstream reservoir (Llyn Clywedog) impoundment. On a much longer timescale, Holocene landforms may still be recovering from the last glacial stage, working to reduce the legacy of glacial sediments that still remain very visible in mid-latitude landscapes. Repeated events may also 'exhaust' the supply of sediment for succeeding events occurring before recovery from the previous one has been able to take place. This means that linear relationships between 'drivers' and 'responses' are replaced by more complex relationships, including what are known as *hysteretic loops* where the relationships between system variables may be in more than one state – for example, river sediment loadings for equivalent discharges on the rising and falling stages of a flood. These may operate both in the short term (as in solute-discharge relationships during a storm), or over seasons and thousands of years, as in the response of sediment yields following yield enhancement during prior glaciation episodes.

Responses may also be absorbed without making the kind of changes that stand out from the variability of event sequences and cycles in naturally variable process histories. Distinguishing 'natural' variability from human-induced change is difficult when there is much event-by-event change taking place, including the incidence of rare extreme events. There are also not one but many interacting and lagged relationships operating in complex geomorphological systems that exhibit sediment storage capabilities, and these may de-synchronize responses, as in catchment sediment outputs where travel times for sediment from different parts of the catchment vary because of variable travel distance and slope (see Jerolmack and Paola, 2010; Van De Wiel and Coulthard, 2010).

14.2 Multiple paths, multiple outcomes and equifinality

Many deterministic numerical models produce single outcomes when change drivers are applied; there is an expected single relationship or pathway from an initial to an end state. Repeat the model runs, and the outcomes are the same – something that even physical modelling seldom achieves with exactness. Huggett (1997a) suggested that more complex 'evolutionary' patterns are in reality more likely because the control factors may themselves be variably interacting and also changing (including crossing thresholds) before any simple response has worked through the system. As shown in Chapter 13, the interlocking allogenic and autogenic forcings can include complex cycles of climatic fluctuation, tectonics that operate to a different spatial distribution and temporal rhythm and, most recently, the globally and temporally varied interventions of anthropogenic factors. The actual changes at any one place are a product not only of the 'global' laws of physics, chemistry and biology, but also of local histories. Their on-going evolution is path dependent, as in other evolutionary models, responding to what has gone before (Chapter 15).

This seems likely to be true of longer-term change trajectories involving multiple nonlinearities, feedback loops and other complexities such as self-organized criticality (SOC) (Chapter 7, Section 7.2). An SOC example is the development of meander loops which evolve progressively but eventually 'go critical' with cut-off formation. Church (1996) has suggested that this should lead to different forms of orderly explanation for different timescales. In the short term, processes are stochastic (individually random but collectively describable in probability terms): deterministic relationships may work at intermediate timescales, whilst at the longest timescales explanations are historically contingent. Here the impact of low frequency high-magnitude events also complicates the picture. Nevertheless, at some scales even chaotic systems (like the movement of bed sediment responding to turbulent flow) may be predictable on average from gravitational forces and generalized pressure gradients.

The outcomes following disturbances may respond to situations where a change of state is possible and they are close to a threshold (Phillips, 2013). There may be a qualitative difference in landform character sometimes called a metamorphosis (Schumm, 1969), thus a channel pattern change from meandering to braiding is more likely if the pattern is in a near-threshold state (Figure 14.3), whilst phenomena like cut-offs or avulsions may be in a prepared state and ready to go in one reach but not the next. The resulting gross channel realignment leads to local upstream and downstream change propagation, as well as the incremental process of loop development by outward expansion or down-river translation. Another way of looking at the nature of adjustments like this is to suggest with Lane and Richards (1997) that 'full explanations' combine *immanent* processes (ones that probably are ahistorical, involving the 'universal' forces of physics and chemistry) and *configurational* factors (which are local and have a history). Thus a successful study of likely future coastal erosion should combine not only quantitative methods for slope stability and sediment transport (often derived from physics and engineering), but also a local configuration and history, unique to the location, and determining boundary and study-start conditions. Field studies are essential to geomorphology as they couple the understanding from process models with the actualities of a local context, and this may often modify the expectations derived from numerical models or controlled laboratory experiments. A reverse corollary is that individual and local field studies may not singularly provide reliable guides to general principles because they are strongly context-dependent.

Another pair of challenging concepts is whether process outcomes (landform sets) are always likely to be varied, or alternatively whether they may converge towards a common end (Phillips, 2013b). The latter is implicit in both the Davisian concept of an 'old age' peneplain, and also in that of characteristic slopes and smooth concave river profiles and end-state entropy maximization. As we saw in Chapter 7, non-linear dynamical systems may become more or less complex with time depending on their scale and histories (Phillips, 1999b). Human activity also leads to variety because it is generally spotty in application (e.g., forest clearance, field cultivation, mining, sea walls, flood embankments or bridges), with processes being locally conditioned by the timing, location and nature of human interventions (e.g., Lewin, 2013).

14.3 Hierarchies and complexities

Landforms are often viewed as being hierarchical in nature (Chapter 4.2); this may involve simply a hierarchy of sizes, as in the framework adopted by Benn and Evans (2010) for classifying glacier erosional forms (**Table 14.4**), and others may be nested in the sense that each level is assembled from entities generated at the level below. This is illustrated by alluvial depositional environments where the base unit may be a single grain of sediment coupled with the drivers generating its mobility: these are combined into elements defined on various grounds (**Table 14.5**) and then into assemblages, variously described as architectural ensembles, fluvial styles, genetic type associations, or allo-formations (Lewin, 2001). The land system concept (and the transferred idea of *glacier* land systems) also addresses the theme of recurrent total landscape patterns, though without necessarily exploring their genetically hierarchical composition. Scientific purposes differ: there may be a need for overall land appraisal, for interpreting the heterogeneous results of long-past deposition episodes, or assembling the multiple process domains through which complex forms emerge. Following Church (1996), spatial as well as temporal scales are influential because larger entities are assemblages, comprising elements and levels developing in different ways. Larger landforms are necessarily more complex to predict than are smaller ones if they combine entities that develop in different ways and rates.

So from a response perspective, process complexity makes change trajectory forecasting more difficult. Thus in alluvial environments, channels may shift at one rate and style, new channels may emerge following avulsions and the cut-off thresholds of different frequencies in pre-prepared locations, while out-of-channel sedimentation may proceed at a different pace driven by extreme flood events and available sediment loadings. Fine sediments are generally present in rivers in sub-capacity quantities, dependent among other things on catchment vegetation cover, human activity and overbank flow resistance and water status. They are thus not easily predictable in field situations using catchment hydrological and channel hydraulic parameters, even though 'reduced complexity' modelling may realistically simulate some essential characteristics of form evolution (e.g., Coulthard et al., 2002). Realistic scenarios may be approached through routing water and sediment across real topographies and using actual climate/vegetation change histories to explore temporal and spatial landform and sediment yield responses (Coulthard and Macklin, 2002; Coulthard et al., 2008). Therefore one can see why Albert Einstein apparently said to his son Hans (a distinguished river engineer) that the field his son worked in was too complex for him!

14.4 Modelling issues

Error is always involved to some degree. Forecasts are always liable to be wrong, despite some analytical outcome producing impressively definite values. *Accuracy* represents the closeness of the forecast to the targeted phenomenon, *precision* the exactness in values achieved. A forecast can be very precise but way out, or it can be in the right zone but only that. Forecasts given to several decimal places are no great help if they are an order of magnitude out, whereas an approximate figure may be quite useful. Predictions of threshold or catastrophic changes, or form metamorphoses, may be particularly valuable. But it is important statistically to define the reliability of predictions wherever possible (their probability levels, or variances). In graphical representations this means plotting percentage confidence limit bands on linear graphs, or producing data spreads in the form of 'box and whisker' plots. In an array of data values, an enclosing box contains 50% of the data and the line 'whisker' the whole range. The conceptual point to be made is that reliability and possible error, as statistical terms, need defining in some way.

Another point is that numerical (or 'software') modelling uses algorithms (step-by-step procedures used in calculation) that frequently rely on applying relationships of the kind outlined in Section 14.1. The name algorithm honours Al-Khwārizmī, a Persian mathematician of long ago (780–850). Such algorithms often use *surrogate models* in their relationships – approximate estimators that are as accurate as possible, as when the forces moving sediment particles are not directly measured but estimated using transport equations regarded as reliable. Approximation is something used very widely in geomorphology, whether estimating erosional activity or fitting lines or surfaces to forms. The use of each approximation involves what is technically described as error, with errors being cumulative in step-by-step procedures, which needs to be stated. Odoni and Lane (2011) have comprehensively reviewed the nature and variability between models used (**Figure 9.1**).

Sensitivity is widely used as a technical term, though with different meanings in different fields of study. Applied in a binary classification test for repeated test result (as in medical trials), it may relate to the proportions of right and wrong test results: the true positive and negative results, and the 'false positive' (Type I error) and 'false negative' (Type II error) results. Sensitivity is the measure of true positive results.

As a concept sensitivity is perhaps most useful in model trials in a different sense: varying inputs or model conditions to assess the degree to which modelling outputs respond to their individual variation. To what environmental factor is change most responsive, or sensitive? This is the kind of experiment for which modelling is well suited: variables can be separately adjusted to explore the results, something that is hardly possible at all in a multivariate world. Thus Coulthard and Macklin (2001) explored the sensitivity of catchments to climatic change, whilst Coulthard and Van de Wiel (2013) modelled the relative impacts of changing climates, tectonics and morphology on sediment yield. They found that significant rates of uplift seemed to get lost in storage, while quite modest rainfall increases were effective in increasing yields. Such results are interesting and suggestive, but they cannot of course be easily checked against a real world situation. However, degrees of sensitivity have been reflected in the use of hypersensitivity – connoting an apparently large adjustment resulting from a small change – and undersensitivity – to denote a disproportionately small response (Brown and Quine, 1999).

How **representative** are the physical or 'hardware' models that have increasingly been used to assess landform development, whether whole but miniature landscapes with 'rivers' and 'hill slopes', or prototype forms like fans, deltas and channels? Some are dimensionally scaled to achieve hydraulic similarity, but others are not. An extensive review (Paola et al., 2009) pointed to the fact that experimental forms organize themselves in such a way as to be remarkably similar to what is observed in the field. Models are 'unreasonably effective', and this suggests that there is a natural scale-independence at work. Such self-similarity indicates that models may be not just miniature analogues or simulations, but also representations at another scale of fundamentals in form development. What this doesn't allow for is absolute time-scaled predictions, but rather analysis of relative or scale-free developmental sequencing. Guidance for planning purposes may be achieved for century-scale change for some forms by examining historical and cartographic data and dated past activity, for example of spit growth. The prediction of futures based on past change trajectories and cycles can be fallible, as economic forecasters and many others have found to their cost.

The greatest difficulty is probably that of testing or **validating** the model match with the real world, not least because the pace and pattern of change are so commonly unobservable directly (a reason why notional models were developed with some enthusiasm from the earliest days of geomorphology). Similar difficulties apply to the validation of future climate change models, and to research by cosmologists who may be able to test some of their predictions, like the position of planets, but who are not (we hope) able to validate models for the ending of the Universe. For rapidly developing forms on Earth, such as gullying, there may be no problem: in other situations some calibration may be achieved by comparing the outcomes of parallel observations or models (e.g., sampled short-term sediment yields with predictions based on distributed source and transport flux outputs), while morphometric properties of modelled forms may be compared with real-world properties (such as drainage densities, network connectivity properties, and measures of channel planform).

Church (2002; 2011), noting that models reduce both time and space scales and simplify conditions – they achieve 'reduced complexity' – suggested three types of model testing or validation: comparison with a 'similar' field situation; comparison with many, so as to explore the variations and differences that might occur anyway; and multiple set comparisons for both trials and landscape sets. The latter works with some models but not with those that are deterministic and achieve the same result every time. Which metric is compared depends on the experiment – it could be drainage network characteristics in catchment modelling or channel pattern in braided stream modelling. Nicholas (2013b) compared ratios of bar width to length for channels, whilst Howard and Henberger (1991) used dimensionless curvatures to compare meanders.

After disturbances, are changes reversible, and will they be repeatable if the same disturbing conditions occur again? This is dependent on the permanence of change, on the temporal spacing of change-driving events in relation to required recovery times, and on other changes that might provide simultaneous environmental transformations – which may mean the configurational context is never the same, so that new disturbances do not take place under the same conditions. **Figure 14.6** shows an interpretation of channel pattern changes on an unstable reach over 150 years, as affected by episodes of larger floods. These were spaced to allow recovery between episodes at c.1880 and c.1960 so that change proved reversible. But this will not now be repeated following reservoir construction in the 1960s 'freezing' the channel pattern in a quasi-braided state so that it has not continued to develop as in previous cycles.

14.5 Conclusion

Geomorphological change trajectories following disturbances are very diffi-cult to forecast. It is feasible for some landform elements where relationships have been established between drivers and responses, as in meander loop development, and where there are changes of state, and where thresholds have been determined. Physical and numerical modelling are providing new insights, though there may be a number of issues concerning what mod-els can do, or as yet can faithfully represent. Field configurations are often operationally unique, especially for larger landforms that comprise multiple element sets evolving in different ways. These configurational contexts also change, not least because of anthropogenic transformations (see Chapter 16).

FURTHER READING

Bishop, P. (2007) Long-term landscape evolution: linking tectonics and sur-face processes, *Earth Surface Processes and Landforms,* 32: 329–65.

Church, M. (1996) Space, time and the mountain – how do we order what we see? In B.L. Rhoads and C.E. Thorn (eds), *The Scientific Nature of Geomorphology.* Chichester: Wiley. pp. 147–77.

Church, M. (2010) The trajectory of geomorphology, *Progress in Physical Geography,* 34: 265–86.

Church, M. (2011) Observations and experiments. In K.J. Gregory and A.S. Goudie (eds), *The SAGE Handbook of Geomorphology*. Thousand Oaks, CA: Sage. pp. 121–41.

Huggett, R.J. (1997) *Environmental Change: The Evolving Ecosphere.* London: Routledge.

Odoni, N.A. and Lane, S.N. (2011) The significance of models in geomorphol-ogy: from concepts to experiments. In K.J. Gregory and A.S. Goudie (eds), *The SAGE Handbook of Geomorphology*. Thousand Oaks, CA: pp. 154–73.

Phillips, J.D. (2009) Changes, perturbations, and responses in geomorphic systems, *Progress in Physical Geography,* 33: 17–30.

Tucker, G.E. and Hancock, G.R. (2010) Modelling landscape evolution, *Earth Surface Processes and Landforms* 35:, 28–50. Doi: 10.1002/esp.1952

Van De Wiel, M.J., Coulthard, T.J., Macklin, M.G. and Lewin, J. (2011) Modelling the response of river systems to environmental change: pro-gress, problems and prospects for palaeo-environmental reconstructions, *Earth-Science Reviews,* 104: 167–85.

1. Why does the hierarchical nature of geomorphological systems pose particular problems for prediction?

2. To what extent can geomorphologists perform experiments?

 WEBSITE

For this chapter the accompanying website **study.sagepub.com/greg oryandlewin** includes Figures 14.2, 14.3, 14.4, 14.6; Tables 14.1, 14.3, 14.4, 14.5; and useful articles in *Progress in Physical Geography*. References for this chapter are included in the reference list on the website.

15

INHERITANCE

Landforms are demonstrably of very different ages; their formation times and their longevity as parts of the landscape are contrasted both globally and within catchments or coastal zones. Available erosional energy (entropy) and resistance are unevenly distributed both spatially and over time, as a result of forcing factor incidence, erodible materials and environmental change. This produces a patchy and overprinted Earth surface on a variety of scales. These inherited forms and materials, interpreted as palimpsests and as patches, continue to influence active process systems.

In a number of senses geomorphology necessarily studies residual land surfaces. In the first place, direct observation of formative activities is often difficult – whether under glaciers, in the sea or in deep rivers, or during extreme and dangerous events that may produce a disproportionate amount of change. We can often also examine only what is left afterwards because many processes operate slowly, with a minimum of centuries to millennia needed for major forms to evolve. Having to 'look later' has produced some interpretations of forms and sediment bodies that now seem quite curious, such as the interpretation of erratics as all being ocean-rafted (though some still are), and even drumlins being related to wind action (Evans, 2008). Where direct observation is not possible, dating techniques become vital for establishing rates and timings – though not of course in any direct way – of formative processes. Not all geomorphological activity is slow; beach profiles and channel beds may undergo multiple changes seasonally, and geomorphology may be involved in examining forms of this kind that interact with flows to give changing platforms for biological activity (Maddock et al., 2013). Similarly the timescale relevant for engineering design may incorporate documented geomorphological changes that are highly relevant for management (Downs and Gregory, 2004).

A second question concerns *formation time* and *duration time*; how long it takes to produce particular landforms, and how long they last without being effectively reworked or destroyed (Figure 15.1). The energy available to do geomorphological work (entropy) is very unevenly distributed, both because of incident process regimes (shorelines, the middle reaches of rivers, and some glaciers being high energy zones), and also because events vary in energy intensity. Rates of process differ in different process domains (**Table 10.2**), and at different times. Resistances also vary, with different vegetation covers and the varied nature of Earth materials that may be lithified, unconsolidated or soluble. On the whole floodplains and beaches, despite their low gradients, are highly dynamic because of localized energy concentrations in wave and river action and erodible material; formation times are rapid, whereas plateaus and interfluves change very slowly. However, landform duration or preservation in high-activity areas (Figure 15.1) may be low because reworking may quickly take place.

In many places there is the question of long-term inheritance: do the contemporary forms observed actually relate to present-day conditions, or are they residual amalgams remaining from past conditions, such as colder climates, higher or lower sea levels, tectonically-driven emergence, or preanthropogenic vegetation covers? Much research in mid-latitudes has been driven to examine the remains of past glaciation by distinguishing and dating glacial phenomena in what are now temperate environments. There

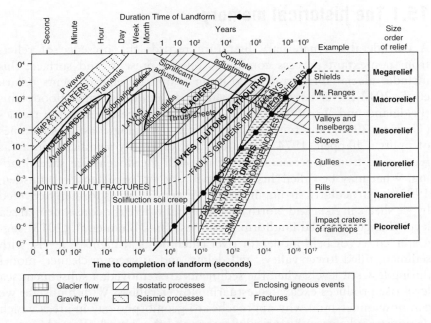

Figure 15.1 Landform formation and duration times (from Brunsden, 1993)

From Zeitschrift für Geomorphologie, D Brunsden (1993) "The Persistence of Landforms" Suppl.-Bd 93, 13–28. Figure 1 (page 18). www.schweizerbart.de

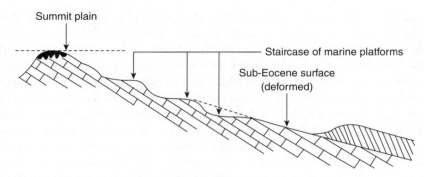

Figure 15.2   An interpretation of planation surfaces in SE England

was also early research on raised beaches, without at first understanding that they were raised as a result of glacio-tectonic effects. The processes currently observable may be greatly conditioned by such past happenings, and regimes now in place would be operating differently if their inherited configurations, both of forms and the pre-weathering of sediments, were different. These questions are addressed in this chapter by memory (15.1), contingency (15.2) and sensitivity (15.3), leading to shape shifts (15.4).

15.1 The historical memory

A considerable Quaternary literature records the presence of relicts from non-extant process conditions on the present land surface; this includes glacial forms and deposits in temperate environments (Clark et al., 2012), river terraces along valleys as elevated former floodplains (Bridgland and Westaway, 2008), palaeoclimate residues in the tropics (Thomas, 2008), and raised shore platforms fringing coastlines in zones of uplift (Chappell, 1974). Land uplift is especially significant in areas recovering from depression caused through Pleistocene ice sheet loading. What may be called 'historical memory' or 'persistence' (Brunsden, 1993) is particularly strong in the very extensive areas formerly covered by ice sheets in Asia and North America, and where the consequential sea level changes affected landform processes. Thus Blum et al. (2013) suggest a 'conveyor belt' model for river catchments at high sea levels, with sediment-filled lower valleys and deltas, and a 'vacuum-cleaner' model during low sea levels when the sediments are removed; a subsequent sea level rise produces excavated estuarine lower valleys. With hindsight we can now understand why approaches to geomorphology in areas which demonstrated a great historical memory and those most affected by contemporary processes have been so contrasted, prompting a request for the re-enchantment of geomorphology (Baker and Twidale, 1991).

Geomorphologists spent a considerable amount of effort in the mid 20th century focused on relict planar forms or 'erosion surfaces' and their reconstruction (Johnson, 1931; Baulig, 1935; Wooldridge and Linton, 1955; Jones, 1980). In England, this research considered a history dating from the Palaeogene and even earlier periods (from c. 50 million years ago): an early and warped 'sub-Eocene surface', an undeformed later summit plain now marked by accordant hilltop levels (Pliocene?, c. 2.5–5.3 million years in age), and a set of lower Pleistocene (?) marine platforms forming a kind of giant staircase at elevations below c. 200m (Figure 15.2). Later research has disputed this picture, pointing to continuing tectonic activity over this long period, the sometimes dubious rationale for correlating fragments of surfaces, which may represent as little as 10% of the contemporary landscape, and the difficulties arising from a lack of dating. Nevertheless some elements of the landscape in this region do seem very long lasting, particularly drainage lines (Gibbard and Lewin, 2003), whilst periglacial landforms also mantle mid-latitude hillsides. There is also an extensive record of cold-stage glaciations and drainage diversions over much of the circum-polar regions (Benn and Evans, 2010). In continental Europe, as well as in North America, there have been many detailed Quaternary studies documenting the interactions between uplift and climatic episodes (e.g., Pazzaglia and Brandon, 2001). A particularly well-studied and well-dated sequence of terraces in the Maas valley in an uplifting area of the southern Netherlands shows development over some 4Ma, with particular terrace levels relating to known Pleistocene climatic cycles (Van Den Berg and Van Hoof, 2001).

Where episodes of different activity styles follow each other, only partially eliminating the forms established by preceding ones, the landscape may appear as a *palimpsest* (**Table 15.1**). The metaphor is that of a re-used slate or parchment where traces of earlier erased text still show through. Thus a valley slope in a now-temperate climate may also have been the wall of a glacier-eroded trough which subsequently received an overlay of periglacial deposits currently being reworked. The slope is a palimpsest.

Globally, as well as within particular catchments, relict landform elements may also be *patchy* in distribution (**Table 15.2**). Long-lasting extensive landscapes are found in cratonic areas like central Australia where there are the remains of drainage systems that pre-date the splitting-up of Gondwanaland about 50 million years ago. By contrast, areas with high contemporary processing rates in actively orogenic and heavily cultivated areas may have few pre-Pleistocene remains. The Hawaiian Islands are a chain of volcanoes ranging in age, from Hawaii with active vents and lava flows, to the extinct Kauaii about 6 million years old. The islands show an age sequence of constructional and then dissection forms. Glacial landforms may similarly show age sequences, with older till surfaces being more dissected, and the extent of inheritance may progressively diminish over time as forms adjust to incident processes.

Within catchments, there may be landform age patchiness on a variety of scales. Ecologists have pointed to floodplain patchiness, some areas of recent geomorphological activity with colonizing flora, and then an age sequence in tree age following the age sequence in landform development. In a classic paper, Everitt (1968) used tree age – as judged by coring cottonwood trees and counting growth rings – as a measure of river channel migration. On the very largest river floodplains like the Amazon, there may also be long-lasting flood basins, resulting from subsidence, that are only slowly infilling adjacent to active rivers which transport sediment and form islands on a very much shorter timescale (Lewin and Ashworth, 2014). Patchiness is also evident through human activity, cultivated fields alongside uncleared forest and urban impervious areas alongside rural land, so that 'pristine' pre-anthropogenic forms may survive only in places.

These two concepts – the palimpsest and the patch – are slightly different in that the former implies an overlaying of later on previous activity such that former conditions still show through, while the latter suggests unequal spatial activity such that some relatively 'old' land surface segments survive to reflect conditions now largely past. The second half of the 20th century witnessed something of a battle between those geomorphologists who saw landform inheritance effects as dominant, and others who thought otherwise. Each group used a different set of analytical tools: dating methods and sedimentary stratigraphy on the one hand, and hydraulics and mechanics on the other. This division seems unfortunate though perhaps professionally inevitable because the technical demands of both groups can be independently all absorbing. Landforms likely to change in years or decades have also demanded attention from a practical management point of view, so that attention has shifted away from some of the earlier questions concerning landforms of considerable longevity. But conceptual conflict becomes unnecessary when it is recognized that long-term persistence does patchily exist and that process analysis, currently focused on high activity zones, has to allow for response and relaxation over a very considerable length of time in some circumstances (Brunsden, 1993; Church, 1996). In many branches of geomorphology it is still not easy to separate products of the past from contemporary systems, for example the role of inheritance of morphology is identified as one of the big problems facing future karst specialists (Ford and Williams, 2011). A further reaction to the implications of inheritance has been the focus on paraglacial landscapes with the term paraglacial defined (Church and Ryder, 1972) as an environment in which non-glacial processes are conditioned by prior glaciations. The paraglacial has now been recognized as a period of readjustment from glacial to nonglacial conditions (e.g., Ballantyne, 2002b; Slaymaker and Kelly, 2007).

15.2 Historical contingency

It may be suggested that, for landforms in an equilibrium state, history is irrelevant (Chapter 2). Forms may vary as events occur, but feedback effects return them to their 'characteristic' condition; that condition may of course be varied because of configurational circumstances like rock type, relief, or vegetation cover. However, if there is a trajectory to development (where the equilibrium is itself changing over time, or there has been no equilibrium established), what happens next may also reflect what has gone before. This is historical contingency (e.g., Phillips, 2001).

With Quaternary climatic change and anthropogenic activities, forms and sediments are present across landscapes that continue to condition what happens: on the one hand this may be identified as materials contingency, or process boundary conditions; on the other, process responses to earlier conditions may also still be in progress. An example might be catchment urbanization or the input of polluted sediment; what happens geomorphologically depends on what was imposed and when, since systems take time to consume – or adjust to – such effects. If the drivers are also on the change (they are themselves dynamic), then the history of that change becomes significant, whether a decrease in energy as relief is lowered, or because of other transformation changes in weathering or incident events or thresholds. The actual sequences of driven changes may impose more or less contingency. Little long-lasting historical contingency is evident when recovery is relatively rapid, but where there are multiple changes that prevent return to stability, then transient non-equilibrium conditions dominate entirely (Figure 14.5).

It could be argued that geomorphological processes today are not nearly as stable as for much of earlier Earth history, the result of climate changes and human activity. Things may now be more historically contingent, not less, despite the popular assumption that we can 'control' our environment to a considerable extent. An example might be the landforms of Appalachian valleys, with fine flood-deposited alluvium capping coarser materials laid down as stream bed-material as rivers migrated laterally. But the coarse material may be a cold-climate inheritance, whilst the exceptional thickness of fine sediment has been shown to date from accelerated soil erosion trapped behind valley-wide mill dams (Walter and Merritts, 2008). It is intriguing that results from a classic study of stream channels of the Watts branch affected by urbanization (Leopold, 1973) may have to be reconsidered as a result of these conclusions about sediment accumulated behind mill dams.

15.3 Geomorphological sensitivity

This term has been used in geomorphology in several ways (e.g., Downs and Gregory, 1995) that are different from the modelling run sensitivity discussed earlier in Chapter 14.4. An important one is that 'The likelihood that a given change in the controls of a system will produce a sensible, recognisable and persistent response' (Thornes and Brunsden, 1979: 476). Brunsden and Thornes use the analogy of a particle in a river with bed hollows and other particles around it; although the particle can be transported by the flow, it will need to get out of the hollow or away from the protection provided by the other particles. The concept is closely related to that of thresholds – as for sediment movement, or to changes of morphological state like river channel patterns (**Table 15.3**). There is therefore a threshold before movement takes place and below which the particle is 'insensitive' to the forces applied. Sensitivity may be with respect to climatic, tectonic or anthropogenic drivers. D'Arcy and Whittaker (2014) have, interestingly, explored the sensitivity of landform parameters to the coupled effects of tectonic uplift and higher precipitation, with the latter suppressing some of the effects of the former. It is often difficult to sort out sensitivities to one or other factors when their effects are operating jointly.

Another idea is that some parts of the landscape are more sensitive than others; there are fast responding sub-environments like river channels and their margins, and others like interfluves that have very little sensitivity at all, partly because of the materials involved (solid rock), but more generally because of a lack of local erosion energy. This is where the long-term preservation of forms is most likely, and why landscapes consist of different age patches. The Brunsden/Thornes analogy was carried further by

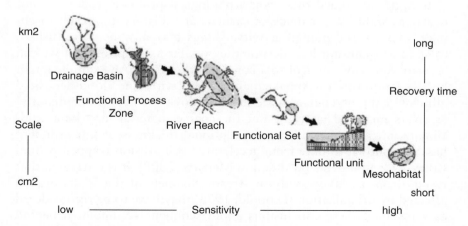

Figure 15.3   Sensitivity in the hierarchy of catchment landforms (after Meitzen et al., 2013)

Fryirs et al. (2007) who suggested that catchment-scale sediment cascades were impeded in their connectivity by 'buffers, barriers and blankets', storages which functioned to disrupt the passage of sediments downstream. It is a matter of opinion as to whether these descriptive analogue terms are really appropriate: is there really a barrier physically in place that actually stops movement? Another way of approaching this is to recognize the spatially distributed nature of processes and process rates, with some being 'slow tempo' zones and storages acting as temporary 'freezing' zones (see Richards and Clifford, 2011) although all are still part of a sediment cascade.

Within catchment systems this leads to different degrees of sensitivity (Figure 15.3). Sediment patches on a channel bed may react rapidly but, as the geomorphic scale increases, the sensitivity decreases so that it takes both more time and many formative events for the larger catchment elements to be transformed, especially those that are remote from concentrated river flows. This may be expressed in terms of a force-resistance balance, with longer-term preservation potential or resilience depending on reworking likelihood (**Table 15.4**).

15.4 Shape shifts: the gradual and the catastrophic

Whilst catastrophism died a death as a general geomorphological framework (Chapter 3), what is sometimes called 'neocatastrophism' remains a concept of some significance in varied form. Extreme events may change landforms rapidly: avalanches, volcanic eruptions, landslides, rock falls, slope instability following earthquake shaking, tsunami effects, or ice-margin effects involving sudden water drainage (Jökulhlaups, large palaeo-lake drainage, and river diversions). These were summarized in Table 3.2, and they also constitute natural hazards (Chapter 17), but these sudden 'shape shifts' are also important components integrated into some landform interpretation models.

For example, in tectonically active landscapes, episodic landsliding may be an important, if irregular, contributor to slope evolution. This may arise through uplift and slope steepening crossing a stability threshold, aided by tectonic shaking and extreme events. Slope development can be driven by rare but collectively dominant events. In some steepland environments, channel headwater gullies may accumulate material slowly over extended periods, releasing and evacuating such fills when a stability threshold is exceeded, probably in an extreme event. In both these cases there may be a cycle of relative quiescence followed by sudden morphological change – both being parts of the same dynamic equilibrium system (Montgomery,

2001). The perspective of the engineer and the geomorphologist may differ in such studies, though not necessarily their methods. Engineers may focus on slope stability to avoid or stabilize such activity, whereas geomorphologists may focus on the totality of landscape after such events had taken place, possibly repeatedly. A final example of 'catastrophic stripping' applies to floodplains. Nanson (1986) suggested that floodplains in southeastern Australia were built up over hundreds or thousands of years and were then stripped away by one or more extreme floods. Lands around the Pacific are liable to cyclical variations in climate with high flood-frequency episodes (the so-called El Niño-Southern Oscillation or ENSO phases), alternating with periods of lower frequency events; and for landform development in such areas these may be regarded as 'normal' components of geomorphological development.

15.5 Conclusion

A traveller to the atmosphere-deficient Moon or Mars a billion years ago would have seen a landscape little different from the present; volcanic resurfacing of Venus seems to have changed its surface entirely in 700 million years (Melosh, 2011). The Earth comes somewhere in between, with active resurfacing because of tectonics and atmospheric and surface reprocessing, but with the continuing preservation of older surfaces. These are unevenly distributed, and landform formation and duration times vary as they respond to active process systems. Landforms may be combinations of different elements – either separate patches of different age and origin thus recording landform history, or combined together and overprinted in complex ways as palimpsests. Some elements may have formed under very different conditions from those of the present, whereas others show inheritance effects only over much shorter timescales, as in beach profiles. Present processes may also *respond* to this mixed configurational heritage that includes the effects of Quaternary climate change and extreme-event episodes, and there are sudden shape-shift thresholds, crossed only rarely, as in the case of mega-landslides.

FURTHER READING

Blum, M., Martin, J., Millikin, K. and Garvin, M. (2013) Paleovalley systems: insights from Quaternary analogs and experiments, *Earth-Science Reviews*, 116: 128–69.

Brunsden, D. (1993) The persistence of landforms, *Zeitschrift für Geomorphologie*, Suppl-Bd. 93: 13–28.

Church, M. (1996) Space, time and the mountain – how do we order what we see? In B.L. Rhoads and C.E. Thorn (eds), *The Scientific Nature of Geomorphology.* Chichester: Wiley. pp. 147–77.

Downs, P.W. and Gregory, K.J. (1995) The sensitivity of river channels to adjustment, *The Professional Geographer,* 47: 168–75.

Melosh, H.J. (2011) *Planetary Surface Processes*. Cambridge: Cambridge University Press.

Schumm, S.A. and Chorley, R.J. (1964) The fall of Threatening Rock, *American Journal of Science,* 262: 1041–54.

TOPICS

1. Can you identify, for an area well known to you, which elements of the landscape are inherited and which result from the operation of current processes?

2. Identify the rare events that are likely to be instrumental in changing the landforms in different Earth environments.

 WEBSITE

For this chapter the accompanying website **study.sagepub.com/greg oryandlewin** includes Tables 15.1, 15.2, 15.3, 15.4; and useful articles in *Progress in Physical Geography*. References for this chapter are included in the reference list on the website.

16

THE 'ANTHROPOCENE'

Whereas geomorphology (landform science) gave insufficient attention to human impact until the second half of the 20th century, subsequent studies of the impact of human agency on the Earth's surface and its processes are now known to be substantial. This has produced a range of overarching concepts and terms. For geomorphologists it has involved measuring the consequences, deciding what is 'natural', and reviewing their research agenda. The degree of human impact has encouraged the proposal that the Anthropocene should be defined as a new era in the geological timescale.

Paul J. Crutzen (Crutzen and Stoermer, 2000) wrote that 'Considering [these and] many other major and still growing impacts of human activities on earth and atmosphere, and at all, including global, scales, it seems to us more than appropriate to emphasize the central role of mankind in geology and ecology by proposing to use the term "anthropocene" for the current geological epoch'. He later commented (Crutzen, 2002) that 'for the past three centuries, the effects of humans on the global environment have escalated. Because of these anthropogenic emissions of carbon dioxide, global climate may depart significantly from natural behaviour for many millennia to come'. These statements were the conclusion to research developments over the previous 50 years. Recent discussions have centred around atmospheric changes, but there have been long-standing changes involving landforms as well. After considering how the impact of human activity was recognized (16.1), the concepts that were associated with an appreciation of the importance of human impacts (16.2), and the implications that arose (16.3), it is possible to consider the justification for the Anthropocene as a new geological epoch (16.4).

16.1 Recognition of human activity

Human activity was not really recognized explicitly by geomorphologists until about 1960 despite many indications that changes had occurred as a result of human action. On reflection it is strange that the whole concept of human activity was so ignored. **Table 16.1** compiles three categories of contributions. First are those in existence prior to 1960, often with significant and insightful conclusions that were insufficiently recognized at the time. It took so long for geomorphology to be affected by these suggestions despite the fact that human activity was becoming increasingly evident from the effects of the dust bowl, the creation of the Tennessee Valley Authority (TVA), and the much greater use of fertilizers, together with the appearance of books on soil erosion (e.g., Jacks and Whyte, 1939). The 1956 book *Man's Role in Changing the Face of the Earth* (**Table 16.1**) did not immediately have the major effect that might have been expected, but it did inform progress subsequently.

Second, in the latter part of the 20th century, contributions established the importance of human activity and provided the foundations for modern research and interpretations. At that time several international and interdisciplinary research programmes reflected increasing environmental concern, including the International Biological Programme (1964–1974), the International Hydrological Decade (1965–1974), which embraced man's influence as one of its major themes, and the Man and the Biosphere Programme (1971–). Environmental concern had mushroomed during the 1960s stimulated by warnings about the impact of human action, by debates about the extent to which Earth resources were finite, and by pessimistic visions of the future for spaceship Earth as exemplified by books such as *Silent Spring* (Carson, 1962). After 1960, geomorphological contributions by Wolman (1967) on the impact of urbanization on sediment yield and channels, and by Vita-Finzi (1969) on the evolution of Mediterranean valleys, underlined the need to focus on human impact. This was reinforced by specific statements by Brown (1970) and Chorley (1971) and then augmented by a series of books (**Table 16.1**), many not exclusively concerned with geomorphology but rather with all Earth systems, culminating in a major volume on *The Earth as Transformed by Human Action* (Turner et al., 1990). Whereas many of the contributions provided exemplars of human impact on particular areas, or in specific branches of geomorphology, it was after 1999 that papers provided analyses of human impact at the global level (Tables 16.2, **16.3**).

Third, examples of the present understanding of the significance of human impact demonstrate how current research is affected. Associated with this final phase, outlined in the third section of **Table 16.1**, have been multidisciplinary ideas which have influenced geomorphology. Thus the 'ecological footprint' is the amount of land required to sustain cities and so expresses their environmental impact. This illustrates how human impact now influences environmental management and decision making (see www.gdrc.org/uem/footprints/what-is-ef.html). Details of the investigation of human impacts are given in **Box 16.1**.

Table 16.2 Comparisons of the efficacy of human agency with 'natural' processes

Rozsa, 2006 (based on data from Nir,1983)

Human activity	Rate of erosion (billion t/yr)
Forest clearing	1
Grazing	50
Land tillage	106
Mining and quarrying	15
Road, railway construction and urban development	1
TOTAL	173

Estimated rates of anthropogenic and natural geomorphological activities (Rozsa, 2006, after Hooke, 1994; 2000; Haff, 2003)

Total estimated rates	Earth moved (billion t/yr)
Anthropogenic	129–134
Natural (including long-distance sediment transfer by rivers 14; meandering by rivers 39; glaciers 4; continental mountain building 14; oceanic mountain building 30; deep sea sedimentation rates 7)	111

In the biosphere ecological impacts on geomorphological processes can be either stabilizing, when vegetation growth reduces erosion, or destabilizing, for example as animal burrowing enhances erosion (Viles et al., 2008). Significant advances made in both biogeomorphology (Viles, 1988; Viles et al., 2008) and geoecology provide new ways of thinking for the whole environmental complex relating to the ways in which animals, plants and soils interact with one another (Huggett, 1995). This then affects geomorphological processes. Although covering just 2% of the Earth's land surface, urban areas are where landforms are largely if not completely the result of human action and where processes have been dramatically changed (Chapter 17).

16.2 Concepts associated with human impacts

As a consequence of the above changes about 75% of Earth's subaerial surface now bears the imprint of human agency, and new conceptual identities have developed. Perhaps the earliest was the addition of *noosphere* to the other spheres that provide the framework for environmental analysis. Vernadsky (1863–1945), a Russian mineralogist and geochemist, is credited with the suggestion (Vernadsky, 1924) that the **noosphere** is the third stage in the Earth's development, succeeding the geosphere (inanimate matter) and the biosphere (biological life). It has therefore become associated with the role of human consciousness in nature – the 'thinking layer' arising from the transformation of the biosphere under the influence of human activity. In that regard Pierre Teilhard de Chardin (1881–1955), a French philosopher and Jesuit priest who trained as a geologist, envisaged the noosphere as emerging through, and constituted by, the interaction of human minds, proposing that it was growing towards an even greater integration and unification. He saw this as culminating at an Omega Point, in the future pulling all of creation towards it, and which he also saw as the goal of history. Osterkamp and Hupp (1996) suggested that de Chardin, by combining evolution with the Greek tradition regarding nature as an organism, was comparing the ontogeny of human society to species whose evolutionary histories necessarily direct their destinies. Subsequently the noosphere was advocated (Trusov, 1969) as a new geological epoch initiated by human activity but the term did not find widespread acceptance (Bird, 1989). Whether such proposals could have scientific evidence to back them is perhaps somewhat debatable.

Subsequent concepts have focused upon the discipline, the sphere, or the division of geological time. *Anthropogeomorphology* was suggested as a discipline devoted to the study of humans as geomorphological agents (Golomb and Eder, 1964; Fels, 1965) and more recently *Neogeomorphology* has been suggested by Haff (2003) as the study of the 'Anthropic Force and its present and likely future effects on the landscape'. He conceived the **Anthropic Force** as including consciousness, intention and design, and produced by the combination of physical and social forces that drive modern landscape change. This includes non-classical geomorphic phenomena, such as landscape planning, engineering, and management. Haff suggested that 'the occurrence of short time-scale phenomena induced by anthropic landscape change, the direct effects of this change on society, and the ability to anticipate and intentionally influence the future trajectory of the global landscape underscore the importance of prediction in a neogeomorphic world'. The human sphere has also been categorized as the *Anthroposphere*, as the *Homosphere* (Svoboda, 1999), which is the biosphere modified by *homo sapiens*, from which the noosphere or 'thinking layer' emanates. Although the *Technosphere* has sometimes been used synonymously with the Anthroposphere, it is usually associated with technical and medical developments, with the online digital environment, or occasionally as that part of the physical environment affected through building or modification by humans. Although Antonio Stoppani (1824–1891), a priest-geologist from Italy, had suggested *Anthropozoic Era* because of the increasing power and impact of humanity on the Earth's systems, and the *Noöcene* epoch was suggested (Samson and Pitt, 1999) to connote how we manage and adapt to the immense amount of knowledge created, these were not widely adopted. It is the relatively recent term Anthropocene that has attracted the most recent interest (Section 16.4 below). It is included with other terms in **Table 16.4**.

One can argue that the wealth of terms outlined above forms alternate namings for reframing cognitive reactions to a common though diverse evidence base for human impacts and transformations (including things like cold-climate thermokarst, accelerated erosion, and landform engineering), coupling some of them with philosophies of human consciousness. Names are of course significant – they define entities – but whether this improves the recognition of human agency, or the evidence base itself, is debatable. From a geomorphological perspective, what is especially required is analysis of the several impacts (Gregory and Goudie, 2011a) with their separable timing and sequencing, so as to establish how landform transformations have occurred and how to deal with them.

16.3 Implications of human agency for geomorphology

The relatively recent acknowledgement of the impact of human agency in geomorphology, and the realisation of its magnitude now equalling or exceeding the results of natural forces, inevitably leads to a series of implications for research in geomorphology and for the way in which the discipline develops. Such implications can be further considered in three related groups: the measurement of human impact (16.3.1); deciding what is 'natural' (16.3.2); and the implications for the future of geomorphology and its research agenda (16.3.3). Hazard study implications are reviewed in Chapter 17.

16.3.1 The measurement of human impact

Measurement of the impact of human agency has shown that the rate of geomorphic change resulting from natural phenomena has now been outstripped by human activities associated with agriculture, construction and mining so that humans are now the most important geomorphic agent on the planet's surface, showing that the accumulation of post settlement alluvium (PSA) on higher-order tributary channels and floodplains (mean rate ~12,600 m/m.y.) is the most important geomorphic process in terms of the erosion and deposition of sediment, far exceeding even the impact of Pleistocene continental glaciers or the current impact of alpine erosion by glacial and/or fluvial processes (Wilkinson and McElroy, 2007). The suspended loads of many rivers have changed significantly in recent years, both increasing and decreasing, and the trajectories of human impact vary markedly (Walling, 2006). The totality of anthropogenic sediment (including marine as well as fluvial material) is also known as legacy sediment (LS) (James, 2013), whilst the term anthropogenic alluvium (AA) may also be used with the implication that more than settlement may be involved for particular impacts, including mining and industrial activity.

New methods of measuring human impact have been devised including *historical range of variability* (HRV) (Wohl, 2011) to describe conditions prior to intensive human alteration of the ecosystem. Often it is not easy to decide what is natural and what is the effect of human agency, and new subjects of attention arise – for example, the cow as an important agent of geomorphological change (Trimble and Mendel, 1995). Desertification, gully development and increasing flood hazard are cases where human activity can be responsible but which could arise from natural processes. The behaviour of Watts Branch, Maryland, was studied over many years showing the consequences of urbanization (Leopold, 1973; 2004), but it was subsequently demonstrated that some of the changes reported could have been the consequence of early dams across the channel (Walter and Merritts, 2008). Given global climate change, is it actually the paramount anthropogenically induced challenge? Slaymaker et al. (2009) contend that 'geomorphologists need, however, to assess whether future climate change is the most important factor in global change or whether other factors (such as land cover change brought about by deforestation) are of greater significance'.

16.3.2 Deciding what is 'natural'

Deciding what is 'natural' is important when landscape restoration is undertaken because it is necessary to decide to what state it should be restored (Chapter 18). Thus Wohl and Merritts (2007) question what is a natural river concluding that, if restoration is designed to facilitate natural river form and process, then it becomes critical to understand how human activities have altered rivers, thereby defining the limits within which restoration can be undertaken. The issue of what is natural is a subject that has engaged philosophers, architects and landscape designers, as well as environmental scientists and geomorphologists.

16.3.3 Implications for the discipline

Implications for the discipline have arisen in three ways. First, in identifying systems according to human agency, such as the 18 anthropogenic biomes distinguished by Ellis and Ramankutty (2008), as a basis for integrating human and ecological systems. Second, in the use of specific concepts or the development of new ones so that **landscape fluidity** introduced in biogeography as the ebb and flow of different organisms within a landscape through time (Manning et al., 2009) could be employed in geomorphology. Third, more general implications include the way in which a more holistic view is encouraged, and the recognition that, although the prediction of human impacts is a compelling need, because of unpredictability many previous generalizations are either wrong or must be adapted to local, regional or environmental-specific conditions (Phillips, 2001). In conceptualizing anthropogenic change Urban (2002) showed how a framework in the assessment of a physical system severely affected by human agency can be applied to the upper Embarras River in east central Illinois that has been physically affected by the cultural practice of agricultural drainage over the past one hundred and fifty years. Changes to the way in which the discipline is viewed with knowledge of how impacts occur has fostered greater knowledge of engineering methods and their significance for processes, thus encouraging more informed applications and the development of branches such as environmental geomorphology (Coates, 1971) and geomorphic engineering. 'Culture' should be accorded greater recognition within geomorphology (e.g., Gregory, 2006), leading to work within wider ecologies of knowledge production (Tadaki et al., 2012).

16.4 The Anthropocene as a new geological epoch

As so much of the Earth's surface has been affected by human agency, the *Anthropocene* was suggested in 2000 by Paul Crutzen, a Nobel Prize winner, because he regarded human impact on the Earth as so significant that it justified naming a new geological era. He proposed this bearing in mind the daunting task for scientists and engineers having to guide society towards environmentally sustainable management during the era of the Anthropocene. Zalasiewicz (et al., 2008; 2011) whose 2008 article is titled '*Are we now living in the Anthropocene?*', argues that our planet no longer functions in the way that it once did, and now is outside Holocene norms. He takes this to suggest that an epoch boundary has been crossed and that there are three possibilities for when the Anthropocene began:

- Thousands of years ago with the rise of agriculture.
- Around 1800 when the human population hit one billion and carbon dioxide started to significantly rise due to the burning of fossil fuels in the Industrial Revolution.

- The end of the Second World War in 1945 when the really big changes were inaugurated.

The Anthropocene proposal gained considerable momentum but whether there is a geomorphological case is reviewed by Brown et al. (2013) debating whether such a subdivision of Earth time is warranted in *geomorphological* terms. The two considerations used are (1) have Earth-surface (geomorphic) process domains (patterns and rates) changed fundamentally during recent Earth history with rising human population? And (2), if so will the consequences of these process-regime changes be preserved in the geological record? In addition of course the near global existence of man-made landforms may now be the distinguishing factor. Impacts are diachronous, and some areas of the world have relatively few so that, if the Anthropocene is adopted, its formal chronological status as an Epoch or Series (the divisions fall into Eras, Periods, Epochs, and Series) is not uncontroversial.

One of the difficulties is that geological divisions must have their boundaries formally defined and their character established at a located stratotype (a Global Stratotype Section and Point, or GSSP); proposals are debated and approved by the International Commission on Stratigraphy. Divisions do change as a result so that, for example, the base of the Quaternary has recently been extended back to 2.58 million years (Gibbard et al., 2009). Other proposals for subdividing the Holocene have also been made, based essentially on climatic criteria (Walker et al., 2012). Again, while geological type-sites are usually in rock, that for the Holocene is an ice core from Greenland. However, for the possible Anthropocene, where would the boundary come (for phenomena that vary in their nature and timing across the globe), and what would the type-site be? A boundary and date might fit one part of the world but not another, a relatively late onset might mean that many human effects that geomorphologists recognize would come outside the Anthropocene, whereas an early date might down-play the drastic effects that have taken place well within the last century (Lewin and Macklin, 2013). Formal status itself does not of course validate the concept nor indeed should it obscure the complex and overlapping nature of human effects.

16.5 Conclusion

Until about 1960 landform science ignored human impact and concentrated on the landforms and processes of those areas not heavily affected by human action. That has now all changed and we accept that we have to investigate such areas – the ones that pose the greatest problems and the ones where we can offer possible solutions. The Anthropocene has become the title for new scientific journals; whatever the final and formal decision about the status of Anthropocene as a new unit in the geological time scale, a useful debate has centred on when it should start. There has also been some thought prompted by when it might end! The title of the most recent book on Gaia, *The Vanishing Face of Gaia: A Final Warning* (Lovelock, 2009), indicates the possible outcome if care is not taken over future human activity and this could involve *A Rough Ride to the Future* (Lovelock, 2014).

FURTHER READING

Brown, A.G., Tooth, S., Chiverell, R.C., Rose, J., Thomas, D.S.G., Wainwright, J., Bullard, J.E., Thorndycraft, V. R., Aalto, R. and Downs, P. (2013) The Anthropocene: is there a geomorphological case?, *Earth Surface Processes and Landforms*, 38: 431–34.

Gibbard, P.L., Head, M.J., Walker, M.J.C., and the Subcommission on Quaternary Stratigraphy (2009) Formal ratification of the Quaternary System/Period and the Pleistocene series/Epoch with a base at 2.58MA, *Journal of Quaternary Science*, 25: 96–102.

Goudie, A.S. and Viles, H. (1997) *The Earth Transformed*. Oxford: Blackwell.

Gregory, K.J. (2006) The human role in changing river channels, *Geomorphology*, 79: 172–91.

James, L.A. (2013) Legacy sediment: definitions and processes of episodically produced anthropogenic sediment, *Anthropocene*. Doi: 10.1016/j. ancene.2013.04.001

Oldfield, F., Barnosky, A.D., Dearing, J., Fischer-Kowalski, M., McNeill, J., Steffen, W. and Zalasiewicz, J. (2014) *The Anthropocene Review*: Its significance, implications and the rationale for a new transdisciplinary journal. *The Anthropocene Review*, 1: 3–7.

Phillips, J.D. (2001) Human impacts on the environment and the primacy of place, *Physical Geography*, 22: 321–32.

Szabo, J., David, LK. and Loczy, D. (eds) (2010) *Anthropogenic Geomorphology: A Guide to Man-Made Landforms*. Dordrecht: Springer.

Walker, M.J.C., Berkelhammer, M., Björk, S., Cwynar, L.C., Fisher, D.A., Long, A.J., Lowe, J.J., Newnham, R.M., Rasmussen, S.O. and Weiss, H. (2012) Formal subdivision of the Holocene Series/Epoch: a discussion paper by a working group of INTIMATE (Integration of ice-core, marine and terrestrial records) and the Subcommission on Quaternary stratigraphy, *Journal of Quaternary Science*, 27: 649–59.

TOPICS

1. Think of other examples that could be included in Table 16.4.

2. Complete examples of man-made landforms. Is this necessary and instructive to identify?

3. Evaluate the case for and against introducing the Anthropocene as a new element of the geological timescale.

 WEBSITE

For this chapter the accompanying website **study.sagepub.com/greg oryandlewin** includes Tables 16.1, 16.3, 16.4; Box 16.1; and useful articles in *Progress in Physical Geography*. References for this chapter are included in the reference list on the website.

SECTION D

DRIVERS FOR THE FUTURE

17

GEOMORPHIC HAZARDS

Hazards have attracted increasing interest over recent decades and the concept of geomorphic hazards embraces all those natural and techno-logical hazards that impact on the Earth's surface, often inducing changes of morphology. Investigations have focused on individual hazards but it is also possible to envisage the way in which a combination of hazards can contribute to the hazardousness of a place. Geomorphic contributions have mapped and modelled hazards, analysed vulnerability, hazard and risk, and suggested management options including those for the prevention of natural disasters. Particular studies have been made in drylands includ-ing desertification, as well as in urban areas. Future potential includes improvement in our understanding of geomorphic hazards by research into the characteristics of hazard events and predicting their occurrence espe-cially as affected by global climate change.

World headlines often demonstrate the sensitivity of the Earth's surface to hazards, but two in 2013 dramatically illustrate the geomorphologi-cal relevance: the *Russian city of Samara being 'eaten alive' by sink-holes* appeared in *The Telegraph* on the 13th of April 2013, and *Indians question how far flash-flooding disaster was manmade* featured in *The Guardian* on the 28th of June 2013. The first, relating to the area at the confluence of the Samara and Volga rivers about 700 miles south-east of Moscow, describes sinkhole-like features which resulted from the thaw-ing of ice in floodplain sediments during the spring thaw. The second occurred in Uttarakhand when a severe cloudburst led to the failure of a glacier, triggering mass movements onto the Hindu pilgrimage town of Kedarnath, with devastating floods, so that the official death toll was 842 by the 29th of June 2013 although many more people were miss-ing. Such disasters (**Box 17.1**), arising from hazards, not only demon-strate the fragility of the Earth's surface environment but also reflect how the perception of surface processes and hazards affects reporting in the media.

Visual information for many other extreme events is available from The Earth Observatory's mission (http://earthobservatory.nasa.gov/

NaturalHazards). National and international awareness of hazards led to the International Decade for Natural Disaster Reduction (IDNDR), designed by the UN in the 1990s to focus the attention of scientists, engineers and economists on the increasing losses and deaths occurring globally as a result of natural hazards. Geomorphologists have considerable relevant expertise (Rosenfeld, 1994) in process studies, the mapping of precursors and antecedent conditions of surficial phenomena, together with understanding of 'magnitude and frequency' concepts, so that a significant practical contribution of geomorphology is the identification of stable landforms and sites with a low probability of catastrophic or progressive involvement with natural or man-induced processes adverse to human occupancy or use (Rosenfeld, 1994).

Environmental hazards are natural process threats that can damage the land surface or property and take human lives, requiring management to mitigate their effects. Many natural hazards are interrelated, as in the way that earthquakes can cause tsunamis. Building upon investigations of human impact, the focus upon hazards occurred in geomorphology over the last four decades as a consequence of three developments (Gregory, 2000: 187–90) prompting research and some emergence of separate disciplines or sub-disciplines: first, awareness of the impact of extreme events; second, the juxtaposition of investigations of physical environment with studies of their socio-economic relevance; and third, appreciation of the importance of perception arising from the difference between the real world and the way that environment is perceived. Natural hazard research focuses upon the interrelation of geophysical events and human activity so they can provide an attractively novel and imaginative teaching vehicle, though such a focus can sometimes distract from sufficient understanding of the processes responsible. Geohazards include a geological state that may lead to widespread damage or risk (www.ngi.no/en/Geohazards /) so many research centres have been established. These include the Geologic Hazards Science Center at Golden, Colorado, which has programmes on Earthquake Hazards, Landslide Hazards, Geomagnetism, and a Global Seismographic Network (https://geohazards.usgs.gov/).

The concepts involved are approached through the range of hazards that are geomorphic (17.1), and the resulting geomorphological research (17.2), leading to future considerations (17.3).

17.1 The range of geomorphic hazards

Geohazards, as a category of environmental hazards, have become associated with the discipline of Geology. Should there be a distinct category of **geomorphic hazards** or do geomorphological hazards transcend classifications? How do we clarify the spread of definitions that have emerged over recent decades? As interest in the hazard field has intensified over the last half century so have the terms, the classifications, disciplines, subdisciplines and concepts, requiring some standardization of the usage of different terms.

A hazard occurs only in the presence of life so that a natural hazard is a threat from a naturally occurring event that has a negative effect on people, property, or the environment. Very extreme events may, or may not, prove hazardous if no person is seriously affected. A working definition of environmental hazards (Smith, 1996: 16) is 'extreme geophysical events, biological processes and major technological accidents, characterized by concentrated releases of energy or materials which pose a largely unexpected threat to human life and can cause significant damage to goods and the environment'. Natural hazards arise from naturally occurring physical phenomena, whereas technological or man-made hazards are caused by human activity usually occurring in or close to urban settlements. The general classification proposed by the Emergency Events Database (EM-DAT: www.emdat.be/) recognizes five categories of disaster: geophysical, meteorological, hydrological, climatological, and biological. However, a disaster should refer to an actual event that has impacted so that this classification should be of hazards. The largest disasters are referred to as catastrophes requiring significant expenditure in time and money for recovery to take place.

Geohazards, an abbreviation for geological hazards, are conditions, processes or potential events that pose a threat to the health, safety or welfare of a group of citizens, or the functions or economy of a community or larger governmental entity (USGS: see **Table 17.1**). Such categories do not include a geomorphic type, underlining the fact that geomorphic hazards transcend the causative categories. Similarly categories of potentially hazardous environmental agencies or processes (Smith 1996: 18; see Table 1.2, modified after Hewitt and Burton, 1971) have five categories: 1. Atmospheric; 2. Hydrologic; 3. Geologic; 4. Biologic; 5. Technologic, with Category 3 including mass movements, erosion, silting, earthquakes, volcanic eruptions, shifting sands – but no mention of geomorphic hazards. A classification constructed to highlight geomorphic hazards (**Table 17.2**) initially discriminates between the spheres (Gregory, 2010: Table 3.1) from which hazards derive. Some primary spheres have specific associated hazards, so that atmosphere hazards include those arising from magnitude, distribution and intensity of precipitation or from temperature extremes, whereas hazard impact is often registered in another sphere, particularly on the land surface. A distinction is sometimes made between meteorological and climatological hazards, with the former arising from events occasioned by short-lived/small- to meso-scale atmospheric processes, whereas the latter are caused by long-lived/meso- to macro-scale processes ranging from intra-seasonal to multi-decadal climate variability. Also arriving in the atmosphere are NEOs (Near Earth Objects) which can impact on the Earth's surface (**Table 17.2**). In the hydrosphere extreme events associated with water occurrence, movement, and distribution can give tsunamis, coastal floods and storm surges, or be associated with river floods and flash floods. Droughts, characterized by a shortage of water, also pertain to the hydrosphere but impact upon the surface and upon other aspects of the Earth's surface system. In the biosphere biological hazards can arise from the exposure of living organisms to germs and toxic substances, epidemics and insect infestations such as locusts, as well as animal stampedes, fungal diseases, poisonous plants and viral diseases. Hazards associated with the lithosphere arise from tectonic movements including earthquakes, volcanic activity including volcanic eruptions, and salt tectonics, with others arising from solution, mass movement, liquefaction and shrink-swell clays, as well as from human impact through ground water management and mining.

Hazards may therefore arise in one sphere but impact in another with some having effects across several spheres. The sphere of human impact, termed the noosphere or anthroposphere (Chapter 16), includes technological or man-made hazards arising particularly from environmental degradation, pollution or accidents (**Table 17.2**). As most hazards actually impact on the Earth's surface then the sphere in which they are registered should be recognized. The appropriate one (Gregory, 2010: Table 3.1) is the geosphere, effectively a blend of several spheres or a zone of interaction on, or near, the Earth's surface involving the atmosphere, hydrosphere, biosphere, lithosphere, pedosphere and noosphere. The geosphere is the obvious location for geomorphic hazards which are identified as of two types – those that have direct geomorphic consequences such as landslides, and those such as atmospheric hazards which may have geomorphic consequences (Gregory, 2010). Gares et al. (1994) noted that the processes that produce geomorphic hazards are rarely geomorphic in nature, being better regarded as atmospheric or hydrologic. He examined geomorphic hazards in four fields: erosion, mass movement, coastal erosion and fluvial erosion. A geomorphic hazard was defined (Slaymaker, 1994) as resulting 'from any landform change that adversely affects the geomorphic stability of a site and that intersects the human use system with adverse socioeconomic impacts'. This included accelerated soil erosion, desertification, floods, landslides, seismicity, soil salinization, thermokarst erosion and volcanic eruptions. In a comprehensive analysis of the *Geomorphological Hazards of Europe* (Embleton and Embleton-Hamann, 1997) the term geomorphological hazard was interpreted broadly to mean any hazard to people and their economic and social infrastructure caused by natural Earth surface processes, or sometimes by human-induced processes that in most cases involve a change in relief. A later edited volume (Alcántara-Ayala and Goudie, 2010) did not provide a definitive list of geomorphological hazards, but referred to them 'such as floods, landslides, snow avalanches, soil erosion, and others'.

The concept of geomorphic hazards has not been articulated definitively because they can occur throughout the geosphere, do not necessarily involve landform change, but do affect people and their economic and social infrastructure. A definition is suggested in **Table 17.2**. Understanding the concept requires consideration of other concepts that have been associated with extreme events and their impact. A further group of terms referring to the impact that may occur includes **risk** as a measure of the probability of a loss of life, property or productive capacity, whereas **vulnerability** is a characteristic of individuals and groups of people who inhabit a given natural, social and economic space (see **Table 17.1**). The hazards matrix developed by Gilbert White (1974) gave a conceptual framework for analysing natural hazards based on seven criteria, was adapted to responses (Gares et al., 1994), and has subsequently been used as the basis for analysis of disturbance parameters (Phillips, 2009; 2011a). The parameters are identified as magnitude, four related to temporal characteristics (frequency, duration, speed of onset, temporal spacing), and two spatial characteristics (areal extent, spatial dispersion).

Such approaches consider hazards individually whereas an alternative concept considers the hazardousness of a place as the complex of conditions which define the hazardous part of a region's environment (Hewitt and Burton, 1971). This was illustrated by analysing the record for southwestern Ontario, revealing that for a fifty-year period there would be: 1 severe drought; 2 major windstorms; 5 severe snowstorms; 8 severe hurricanes; 10 severe glaze storms; 16 severe floods; 25 severe hailstorms; and 39 tornadoes. Hazards were taken to be simple, which included a single damaging element such as wind, rain, floodwater or Earth tremor; compound, which involves several elements acting together above their respective damage thresholds such as the wind, hail and lightning of a severe storm; and multiple, when elements of different kinds coincide accidentally or follow one another, as a hurricane may be succeeded by landslides and floods.

17.2 Geomorphological contributions

Many geomorphological contributions to the study of hazards have been made by geomorphologists but one, a book entitled *The Time of Darkness* (Blong, 1982) – written about a volcanic eruption in Papua New Guinea – provided a fascinating evaluation of myth and reality by comparing 54 versions of local legends with the scientifically reconstructed Tibito tephra eruption involving a thermal energy production of 10/25th ergs, one of the greatest eruptions of the last thousand years.

Such a study, evaluating perception against reality, demonstrates what can be explored, but most research investigations fall into three main areas (Hewitt, 1983): the monitoring and scientific understanding of geophysical processes; planning and managerial activities to contain the processes where possible; and emergency measures. Several edited volumes demonstrate geomorphological contributions (e.g., Slaymaker, 1994) and Alcántara-Ayala and Goudie (2010) argue that geomorphological work, as well as the mapping and modelling of Earth's surface processes, also includes increasing involvement with the dimensions of societal problem solving, through vulnerability analysis, hazard and risk assessment, and management. The work of geomorphologists can be of prime importance for disaster prevention, for example by documenting hazards in relation to Earth surface processes. Work by the IGU project *Geomorphological Hazards of Europe 1989–1997* aimed to synthesize and analyse information on geomorphological hazards in all European countries as a major European contribution to the International Decade for Natural Disasters Reduction. The results (Embleton and Embleton-Hamann, 1997) provided reviews of what is known about hazards in each specific country. The geomorphological emphasis has been upon hazards arising from the range of surface processes by focusing on the dynamics of the landforms. Problems addressed have ranged from prediction of occurrence, the determination of spatial and temporal characteristics, the impact of physical characteristics on people's perception, and the impact of physical characteristics on adjustment formulation (Gares et al., 1994). Investigations have emerged from specific branches of geomorphology, such as mass movement processes illustrated by geomorphological mapping for hazard assessment in a neotectonic terrain (Petley, 1998), and by volcanic hazards (Blong, 1984), where it has been suggested (Thouret, 1999) that volcanic geomorphology is essential for risk assessment through geomorphic hazard zonation and composite risk zonation, such treatments being necessary to face the enhanced challenge posed by the combination of natural hazards and the increasing number of people who are at risk around volcanoes. Although understanding and reducing vulnerability is undoubtedly the task of multi-disciplinary teams, Alcántara-Ayala (2002: 108) contends that among geoscientists, geomorphologists with a geography background might be best equipped to undertake research related to the prevention of natural disasters, given their understanding not only of the natural processes but also of their interactions with the human system. In this sense, geomorphology has contributed enormously to the understanding and assessment of different natural hazards (such as flooding, landslides, volcanic activity and seismicity) and, to a lesser extent, they have started to consider the field of natural disaster impact.

In addition to thematic ways in which geomorphologists have undertaken hazard research, two spatial areas, drylands and urban areas (both concepts), have merited particular attention. Desertification refers to the degradation of land in drylands, arid, semi-arid, and dry sub-humid areas, caused primarily by human activities and climatic variations (FAO, 2005). This affects 70% of all drylands, and one-quarter of the total land area of the world. Context is provided because the United Nations General Assembly declared 2006 to be the International Year of Deserts and Desertification (IYDD). The worldwide evidence for desertification indicates (Williams and Balling, 1996) that the more obvious signs of dryland degradation include: accelerated soil erosion by wind and water; salt accumulation in the surface horizons of dryland soils; a decline in soil structural stability with an attendant increase in surface crusting and surface runoff, and a concomitant reduction in soil infiltration capacity and soil moisture storage; the replacement of forest or woodland by secondary savanna grassland or scrub; an increase in the flow variability of dryland rivers and streams; an increase in the salt content of previously freshwater lakes, wetlands and rivers; and a reduction in species diversity and plant biomass in dryland ecosystems. It is thought that desertification in China is probably controlled by climate change and geomorphological processes, exacerbated by human impacts (Wang et al., 2008).

Urban areas, although ignored by geomorphologists until relatively recently (Gregory, 2010: Chapter 11), are places where the awareness of hazards is extremely high. Case studies undertaken of specific cities including Los Angeles (Cooke, 1977), characterized by Whittow (1980) as an environment which experiences an accumulation of hazards, were highlighted in *Urban Geomorphology in Drylands* (Cooke et al.,1982). The flood hazard – its relation to urban stormwater provisions, often designed some years ago and possibly outdated by increased frequency of peak flows – is often cited. In addition the spatial distribution of individual urban hazards can be important as demonstrated by problems associated with stream channel change (Table 17.3) identified for Fountain Hills, Arizona (Chin and Gregory, 2005). Many developing countries have rapid urbanization with cities in hazardous or environmentally sensitive areas. This requires the interfacing of geomorphology with engineering practices and urban planning, particularly for cities like Kingston or Bangkok where hazards are so acute and widespread that a practical solution is difficult to achieve (Gupta and Ahmed, 1999). Geomorphologists now involved in research in urban environments often collaborate with other scientists as multidisciplinary investigations are required to respond to legislative requirements such as the Water Framework Directive.

The contrast between perception and reality which introduced this section may affect decision making and environmental management. Three types of erroneous misperceptions for fluvial hazards have been identified (Schumm, 1994): of stability – any change is not natural; of instability – change will not cease; and of excessive response – change is always major. Such misperceptions can lead to litigation and unnecessary engineering works, as well as demonstrating that geomorphic interest should not be confined to analysis of the character of natural geomorphic hazards.

17.3 Future developments

Identifying areas that are vulnerable to hazards and focusing on associated risk are ways in which further research will continue to be required, with opportunities for geomorphologists to improve understanding of geomorphic hazards through research into the characteristics of the events and predicting their occurrence, together with an increased evaluation of the impact of human activities on natural systems (Gares et al., 1994). Additional potential exists for geomorphologists to focus on the changing significance of hazards. Change can arise because of increased human impact, and the natural environment is becoming more hazardous in a number of complex ways because losses are rising and catastrophe potential is enlarging, with costs falling inequitably amongst the nations of the world (Burton et al., 1978). Not only may more people be vulnerable but also hazards are now reported more readily and more rapidly than before communications were so developed: the increase in events reported in the *Annual Register of World Events*, produced since 1757, together with the contents of *The Times*, indexed since 1790, can owe much to the increased reporting of hazards (Gregory, 1992).

Global warming could be one major reason for future change. This concept (Chapters 19 and 20) is thought of as the rise in near-surface atmospheric temperature due to human activities, especially increased amounts of greenhouse gases. Becoming part of a high CO_2 world as a result of the greenhouse effect has consequences for the land surface. The incidence of hazards and natural disasters (e.g., Houghton, 1997: Table 1.1) merits careful scrutiny. Modelling studies and the projection of current trends point towards future increased risk in relation to a spectrum of geological and geomorphological hazards in a world warmed by anthropogenic climate change (McGuire et al., 2010), with observations already suggesting that the on-going rise in global average temperatures may be producing a hazardous response from the geosphere. Evaluative geomorphological studies already provided include: the demonstration that climate warming is influencing permafrost in the Zugspitze area (Bavarian Alps, Germany), with significance for the foundations of existing and future structures (Gude and Barsch, 2005); that increased sedimentation rates in eight estuaries of northern Spain, particularly after the 1950s, may be due to human-induced geomorphic change rather than climate change (also human-induced), as often assumed (Bruschi et al., 2013); and how future human activity and climate change may modify the operation of various geomorphological hazards in Latin America (Goudie, 2009). Potential changes depend upon threshold resilience, whether global climate change is compounded by other more local human actions, the susceptibility of landforms to change, and the severity of climate change. Climate and environmental changes associated with anthropogenic global warming, increasingly identified in the European Alps, affect land surface stability, leading to an increased frequency and magnitude of natural mountain hazards, including rock falls, debris flows, landslides, avalanches and floods, as well as infrastructure, and socio-economic and cultural activities in mountain regions (Keiler et al., 2010). Two case studies (a 2003 heat wave, and floods in 2005) demonstrated some of the inter-linkages between physical processes and human activity in climatically sensitive alpine regions, prompting formulation of a risk cycle related to global warming (**Figure 17.1**) which can be set in the context of the ingredients of natural disasters (Alcántara-Ayala, 2002).

Such promising avenues for geomorphological research need to be cognisant of complexity (see Chapter 7), in the way that the relationship between land-use management and fluvial flood risk can be explored (Pattison and Lane, 2012) using Sayer's (1992) notion of a 'chaotic conception', and also of uncertainty viewed by Brown (2010) as an 'information deficit' to be resolved rather than an inherent product of conducting research.

FURTHER READING

Alcántara-Ayala, I. and Goudie, A.S. (eds) (2010) *Geomorphological Hazards and Disaster Prevention*. Cambridge: Cambridge University Press.

Douglas, I. (2013) Climate change and the ground we build on, *Town and Country Planning*, 80: 426–30.

Gares, P.A., Sherman, D.J. and Nordstrom, K.F. (1994) Geomorphology and natural hazards, *Geomorphology*, 10: 1–18.

McGuire, B., Mason, I. and Kilburn, C. (2002) *Natural Hazards and Environmental Change*. London: Arnold.

Phillips, J.D. (2011) Disturbance and responses in geomorphic systems. In K.J. Gregory and A.S. Goudie (eds), *The SAGE Handbook of Geomorphology*. London: Sage. pp. 555–66.

Slaymaker, O. (ed.) (1994) *Geomorphic Hazards*. Chichester: Wiley.

Smith, K. (1996) *Environmental Hazards.* London and New York: Routledge.

TOPICS

1. Research the Tibito tephra eruptions, the major one in the mid 17th century, over central New Guinea, to illustrate the way in which known fact differs from myth and perception. In addition to Blong (1992) other references include www.jps.auckland.ac.nz/document/ Volume_89_1980/Volume_89,_No._3/Legendary_volcanic_eruptions_and_the_Huli,_Papua_New_Guinea,_by_Bryant_J._Allen_and_ Andrew_W._Wood,_p_341-348/p1 ; Grattan, J. and Torrence R. (eds) (2007) *Living Under the Shadow: The Cultural Impacts of Volcanic Eruptions*. California: Left Coast Press. p. 186; Blong, R.J. (1984) *Volcanic Hazards: A Source Book on the Effects of Eruptions.* Sydney: Academic Press; Cronin, S.J. and Cashman, K.C. (in press) Volcanic oral traditions in hazard assessment and mitigation. In J. Grattan and R. Torrence (eds), *Living under the Shadow: The Archaeological, Cultural and Environmental Impact of Volcanic Eruptions*. London:UCL Press; http://adsabs.harvard.edu/abs/2008AGUFM.V11C2067H

2. The concept of global warming was the title for an article by Keshav Chaturvedi, in the *Hindustan Times* (New Delhi), 4th January 2010. Does global warming stand as a concept? Discover how it has developed since the greenhouse effect was proposed by Joseph Fourier in 1824, discovered in 1860 by John Tyndall, and was first investigated quantitatively by Svante Arrhenius in 1896.

 WEBSITE

For this chapter the accompanying website **study.sagepub.com/ gregoryandlewin** includes Figure 17.1; Tables 17.1, 17.2, 17.3; Box 17.1; and useful articles in *Progress in Physical Geography*. References for this chapter are included in the reference list on the website.

18

GEOMORPHIC ENGINEERING

Concepts of environmental geomorphology and geomorphic engineering were introduced to headline ways in which geomorphological research can be directly relevant to environmental management and planning. They provide a focus for applied research embracing environmental auditing, impacts, evaluation and prediction/design, with many successful applications now demonstrated and core skills recognized. Further potential for geomorphic engineering could reinforce a holistic approach that positions specific problems in their spatial and temporal contexts

In 1976, Coates (1976a: 6) advocated a field of geomorphic engineering as combining 'the talents of the geomorphology and engineering disciplines … The geomorphic engineer is interested in maintaining (and working towards the accomplishment of) the maximum integrity and balance of the total land-water ecosystem as it relates to landforms, surface materials and processes'. This contrasted with engineering styled as 'the art or even the science of using power and materials most effectively in ways that are valuable and necessary to man' (Coates, 1976b). Engineering geomorphology 'complements engineering geology in providing a spatial context for explaining the nature and distribution of particular ground-related problems and resources … concerned with evaluating landform changes for society and the environment' (Fookes and Lee, 2005: 26). It is instructive to elaborate why geomorphic engineering arose and what led to its identification (18.1), how applications have progressed in the last four decades (18.2), and which requirements are outstanding (18.3).

18.1 Why geomorphic engineering?

Understanding the way in which applications of geomorphology developed is necessary to appreciate the 1976 concept of geomorphic engineering. Research has been categorized as *blue skies* research (not specifically related to environment problems and not profitable in the current state of knowledge or technological development), *applicable* or *grey skies* research (investigations which give new results or new facts which may be applicable to environmental problems) or *applied* research (results directly related to environmental problems in a specific area). The development of process studies (Chapter 9) and increasing recognition both of human impact (Chapters 9, 10 and 16) and of hazards (Chapter 17) encouraged greater awareness of the potential of geomorphology to contribute to environmental management so that applicable research increased, with impressive examples of geomorphology being used for consultancy (e.g., Brunsden et al., 1979; 1980). However, such an increase in applicable and then applied geomorphology really began only after 1970; books becoming available were one stimulant for this development. In addition to awareness in geomorphology two other influential trends were developments in related sciences and the context provided by more general initiatives. **Table 18.1** indicates some key developments.

The potential for applicable geomorphology became more evident not only from implications arising from studies of process but also from developments of land classification. The former arose through the appreciation of environmental impacts – not only could processes have an impact but also human activity could affect processes. Relevant legislation included the National Environmental Policy Act (NEPA) in 1969, establishing a US national policy promoting the enhancement of the environment and stimulating procedural requirements for federal government agencies to prepare environmental assessments (EAs) and environmental impact statements (EISs). EAs and EISs contain statements of the environmental effects of proposed federal agency actions. This was succeeded by developments in other countries. In the EU the European Union Directive (85/337/EEC) on Environmental Impact Assessments (known as the *EIA Directive*) was first introduced in 1985 (and subsequently amended in 1997, 2003 and 2009).

Investigations of land classification, not confined to geomorphology, included landscape ecology which described and characterized landscape according to relationships between the biosphere and the anthroposphere, an approach interpreting landscape as supporting inter-related natural and cultural systems (Vink, 1968; 1983). Although the concept was first used by C. Troll, there were several antecedents for the description of the physical environment in a way pertinent to its utilization and management including the recognition of land systems (Chapter 4). Whereas landscape ecology, subsequently a field in its own right involving several disciplines, focused on what land could be used for, land evaluation is the estimation of the potential of land for specific kinds of use, effectively comparing the requirements of land use with the resource potential offered by the environment. Thus Isachenko (1973b) showed how three scales of landscape research for planning purposes could be developed: *evaluation* maps classify terrain for a particular purpose; *prediction* maps indicate the modifications likely to arise; and *recommendation* maps show the measures which could be used to change the environment – this was illustrated by evaluation of land with a reference to tourism (Isachenko, 1973a).

It was against this background that Coates (1971) defined **environmental geomorphology** as 'the practical use of geomorphology for the solution of problems where man wishes to transform landforms or to use and change surficial processes'. Structural solutions applied to the management of the environment in a particular area have come to be known as hard engineering, and research since the 1970s demonstrated how this could have significant effects elsewhere: increased flooding downstream of channelization works; changed coastal erosion patterns adjacent to coast protection measures; sedimentation and channel changes downstream of dams; and thaw features where permafrost has been thermally affected (see Gregory and Walling, 1987). Geomorphic research focused on the less obvious consequences of hard engineering methods, fostering a greater awareness of environmental consequences and for fluvial systems what Leopold (1977) styled a 'reverence for rivers'. Subsequently Coates (1976) suggested that the geomorphologist 'must become involved in the tools of engineering because if construction causes irreparable damage to the land–water ecosystems, due to lack of geomorphic input the Earth scientist cannot be absolved of blame'.

18.2 Progress of applications of geomorphology

The essence of a geomorphic engineering approach is to avoid engineering that does not take sufficient account of the spatial and temporal context; this might lead to problems encapsulated by Coates as 'The operation was a success but the patient died'. It was an idea that was subsequently embraced as part of an holistic approach. Coates appreciated that to designate a specific approach as 'geomorphic engineering' may appear as a heretical act of nomenclatural proliferation but claimed that such a field is vital, although some (e.g., Fookes and Lee, 2006) have preferred 'engineering geomorphology'. More recently the need for a more holistic awareness has been embraced within soft engineering (see Downs and Gregory, 2004).

Applications of geomorphology have been achieved in a context of national and international trends including growing environmental awareness dating from *Silent Spring* (Carson, 1962) and an article on 'The tragedy of the commons' (Hardin, 1968) which drew attention to areas of land

and water that could be regarded as commons but had been brought to the brink of environmental catastrophe by individuals acting quite appropriately but alone and without respect for the common good. Subsequently these were reinforced by public attitudes to nature (e.g., Nash, 1967), awareness of limits to growth (Meadows et al., 1972), green movements (e.g., Frodeman, 1995), and environmental ethics (e.g., Taylor, 1986; Cahn, 1978). Legislation at national and international levels (**Table 18.1**) formalized the framework of environmental concern and awareness including the NEPA in 1969, and the UNWCED report *Our Common Future* in 1987 that defined the nature of 'sustainable development' and advanced seven strategic imperatives and seven preconditions for sustainability to be achieved. Follow-ups included *Caring for the Earth: A Strategy for Sustainable Living* (IUCN, 1991) which based its strategy on nine principles. Since 1992 UNCED international conferences have addressed the challenges of global change including climate change (Rio de Janeiro, 1992), and Agenda 21 – introduced at Rio in 1992 – is a UN action agendum referring to the 21st century for other multilateral organizations and for individual governments around the world that can be executed at local, national, and global levels. The IPCC (Intergovernmental Panel on Climate Change) currently has 195 countries as members, and produced its first assessment report in 1990 with subsequent ones in 1995, 2001, 2007 and 2013. Training provision expanded so that by 1975 at least 43 major universities had instituted graduate degree programmes of environmental engineering (Coates, 1976: 15).

Such developments provide the opportunity for applications of geomorphology (Table 18.2) and these have addressed four tasks. First, the description, depiction and auditing of the environment in a relevant way are fundamental for many specific applications; this requires not only selecting relevant ways of characterizing landforms and processes but also of communicating information in such a way as to be readily understood by planners, with GIS as an appropriate tool for assembling and providing information. At this descriptive or audit stage it is necessary to identify the variables involved in a particular environmental situation, to propose ways of describing the physical environment or the processes operating in a way relevant to the problem in hand, to indicate how data may be obtained, from existing records or by field or remotely sensed survey, and how these can be analysed. Audit of spatial distribution is important in the way that river channelization 1930–1980 for England and Wales (Brookes et al., 1983) provided a benchmark for detailed investigations. Methods are necessary to describe the general context of a specific project site, or the nature of sites for engineering construction of roads, buildings or dams, as well as describing natural hazards. It is then often necessary to audit geomorphological processes, particularly those that might occasion natural hazards.

Table 18.2 Types of geomorphological contributions (developed from Table 12.3, Gregory, 2010)

Type of Contribution	General Examples	Specific Examples
Audit Stage: Description, depiction and auditing of physical environment	Erosion potential of coast Scenic character of landscape Slope erosion classification Land classification Landscape ecology Classification of land according to potential for soil erosion	Flood frequency analysis, including palaeoflood data and compared with analysis based upon continuous records necessary to improve flood frequency distribution and estimations of discharges with RI of say 50 or 100 years, can be radically different for Western Arizona (House and Baker, 2001) Under-design of dam spillways in NSW in analysis of Australia's fluvial systems (Tooth and Nanson, 1995) Channel incision related to gravel mining of channels (Uribelarrea et al., 2003)
Environmental Impact: Investigation and analysis of environmental impacts	Extent of coastal floods Effects of dams on flows and channels downstream Drought impact in particular areas Effects of crop practices on soils Accumulation of metal contaminants in soils Impact on wetlands Impact of pressures on national parks, nature reserves	Locational probability of channel change 1935–1997 could indicate the most probable location and configuration of the channel for urbanizing stream (Graf, 2000) Estimation of channel recovery potential based on connectivity of reaches and limiting factors to recovery (Brooks and Brierley, 2004)
Evaluation of environment and environmental processes: Evaluation of environment to show how certain characteristics are appropriate for a particular form of action	Characterization of avalanche slopes in relation to transport and to settlement location Evaluation of soil capability for agricultural and other land uses Land use (vegetation systems in relation to environmental management)	For hazard analysis and global change examples see Chapter 17 Management options for coarse woody debris in channels (Gurnell et al., 1995) For design applications see Chapter 19
Prediction, design and policy-related issues: Prediction and design concerned with future and policy-related uses	Specific highway location and design according to slope stability Vegetation and textiles to control soil erosion Buffer strips along rivers	

Second, environmental impacts need to be investigated as an integral part of studies of relationships between Earth processes and landforms, with an emphasis on the magnitude of environmental impact already sustained so that future estimates of sustainability can be made. Techniques can be developed for Environmental Impact Assessment (EIA), which is the process of determining and evaluating the positive and negative effects that a proposed action could have on the environment, before a decision is taken on whether to proceed or not. The International Association for Impact Assessment (IAIA) defines an environmental impact assessment as 'the process of identifying, predicting, evaluating and mitigating the biophysical, social, and other relevant effects of development proposals prior to major decisions being taken and commitments made' and many manuals are available (e.g., Petts, 1999). However, as not all the effects that a proposed plan may have on the land surface or on Earth surface processes are easily identified, there are many opportunities for the geomorphologist to assess all potential impacts. This can require reviewing the purpose and need for the proposed action, identifying the environment that could be affected, evaluating the alternatives available, and analysing the impacts of possible alternatives. Recognition of the importance of a sound understanding of geomorphological processes (e.g., Thorne and Thompson, 1995: 583–84) is vital for geomorphic engineering.

Third, evaluation to show how certain characteristics of environment are appropriate for a particular form of utilization develops from methods of land evaluation which estimate the potential of land for specific kinds of productive use, such as arable and other types of farming, forestry, uses for water catchment areas, recreation, tourism and wildlife conservation. This has to take account of the requirements of legislation in the particular country, and also of any judgements or preferences that may reflect the local culture, before proceeding to complete an evaluation for a particular purpose. Research for planning purposes requires evaluation maps classifying terrain for a particular purpose, and prediction maps indicating the modifications likely to arise, with recommendation maps showing the measures which could be used to change the environment (Isachenko, 1973b). Many examples that suggest solutions to specific problems have now been completed (Table 18.2).

Fourth, future – and often policy-related – uses involve prediction and design. Whereas the evaluation of environment is primarily devoted to the uses of contemporary environments, prediction, design and global change impacts are more concerned with the future and associated with the policy issues necessary to influence developments as shown in Chapter 19.

These four broad categories (Table 18.2) embrace types of application of geomorphology, but it has to be remembered that what is viewed as geomorphology in one country may elsewhere be partly, or completely, included within other disciplines. However, the significant progress achieved (see Gregory, 2010: 292) has been shown in at least three major ways.

1. Research and publication – scientific papers and books (**Table 18.1**). An increasing number of research papers include explicit planning or management considerations. In fluvial geomorphology explicit recommendations for management of urban river channels were found in 47 research investigations (Chin and Gregory, 2009: Table 3). There are also many restricted-circulation reports that exemplify contributions by geomorphologists to planning problems.

2. Individual geomorphologists working either in a role influencing policy making in specific environmental organizations, or groupings of scientists specifically founded to undertake consultancy research. When commissioned as consultants, geomorphologists can bring the advantage of an holistic view of the land surface – a perspective not always achieved by other disciplines.

3. Geomorphologists employed professionally using their degree programme geomorphological training to contribute directly towards decision making. Practical application and experience are now recognized for geomorphologists by chartered status for example (www.rgs.org/pdf/CGeogApplication).

Therefore, what are the key features of the concept of applied geomorphology? This term is often used somewhat interchangeably with environmental management or geomorphic engineering, and much progress has been achieved since 1979 when the question was posed as to how applied we should become (Gregory, 1979). Applicable research, focused on the impact of engineering and management practices, was succeeded more recently by research applied to the solution of specific problems. For hazards and extreme events (Chapter 17) the emphasis has been upon hazard avoidance, but other research has necessarily taken account of complexity (Chapter 7), the value of a holistic approach (Chapter 2), the fact that geomorphic change trajectories are difficult to forecast (Chapter 14), sensitivity (Chapter 15), and uncertainty (Chapter 19). Techniques available in recent years have enabled a focus on extending the reference time horizon, employing historical time series and models, longer time series, dating techniques and climate change data, and climate models, together with broadening the spatial analytical context using digital elevation models and remote sensing (Gregory and Downs, 2008). An holistic view relies not only on spatial analytical context but also on adopting a longer-term perspective which has the advantages of avoiding impulsive responses to individual events by viewing them as part of the long-term flow and sediment record, emphasising preventative planning to accommodate uncertain aspects of change, and avoiding costly engineering mistakes where the structural 'solution' is either inappropriate to the problem faced – 'over-engineered' in relation to hazard – or is wrongly positioned (Gregory and Downs, 2008). Recent progress has allowed the core skills of the applied geomorphologist to be suggested (Downs and Booth, 2011), thus providing an alternative and more expansive way (**Table 18.3**) of specifying the tasks of applied geomorphology indicated in Table 18.2.

18.3 Outstanding requirements

Specific examples (final column of Table 18.2) and the skills of the applied geomorphologist (**Table 18.3**) reflect the very significant progress achieved in geomorphic engineering, anticipating design (Chapter 19) and global change (Chapter 16). However, recent research, for example on resistant-boundary mountain channels, shows how process and form are inherently non-linear (Wohl, 2013), with the management implication that a formulaic one-size-fits-all approach is inadequate.

Any former reticence to become involved in applied research (Coates, 1982) has been replaced by a willingness to focus upon specific developmental problems, as aided by enhanced modelling capability, greatly accelerated by technique advancement. The ability to take a holistic view embracing both spatial and temporal contexts has countered the greater specialist emphasis upon components of the land surface without sufficiently acknowledging the links between them (Gregory and Goudie, 2011a: 16–17). Because **holism** applies literally to the whole as more than the sum of the parts, it is the basis for greater links developed in multidisciplinary investigations between geomorphology and other disciplines such as the interface of geomorphology and ecosystems ecology (e.g., Renschler et al., 2007). Hybrid disciplines have been fostered, including ecogeomorphology and hydrogeomorphology, and multidisciplinary investigations have been positively encouraged. In fact, paradoxically, the success of geomorphic engineering applications might lead to the demise of geomorphology as an over-distinct discipline as more multidisciplinary inputs become necessary. Developments have been particularly evident in fluvial geomorphology (Thorndycraft et al., 2008) with a resurgence taking place, fostered for example by an interaction with river engineering and the availability of new analytical methods, instrumentation and techniques. These have enabled the development of new applications in river management, landscape restoration, hazard studies, river history and geoarchaeology. Geomorphologists have overcome their reluctance to consider problems in urban areas (e.g., Gregory, 2010: Chapter 11) where there are opportunities involving a holistic approach, with restoration where feasible, compatible with a sustainable urban future (Chin et al., 2013).

Many of the problems or challenges to be surmounted are founded upon both scientific and intra-society communication, where geomorphology needs to be aware of the changing situation (Gregory et al., 2013) and to use the potential available. The 'paradigm lock' (**Figure 18.1**) expresses the possible gulf between scientists who do not grasp what managers require, and managers and stakeholders who do not appreciate the scientific alternatives available (Bonnell and Askew, 2000; Endreny, 2001; Gregory, 2004a; see also Gregory, 2010). A weakness at the interface between environmental advice and managerial decision making can account for many failures of environmental policy (Cooke, 1992). Research needs to continue to respond to current developments in the way that the Water Framework Directive (Directive 2000/60/EC of the European Parliament and of the Council of 23 October 2000 that established a framework for European Community action in the field of water policy) provided opportunities for applied research. There is clearly a need to make geomorphology more visible (Tooth, 2009) and to ensure that the results of research are communicated, not only including publication outputs embracing the range from review papers, book chapters, and books, but also applied outputs which cover interdisciplinary problem solving, educational outreach, protocols and direct involvement (Gregory et al., 2008) in suitable formats.

Recognizing such difficulties is imperative but considerable potential remains, as indicated by Graf (1992), who contended that effective science and well-informed public policy are the avenues to managing environmental resources successfully. In reviewing geomorphological involvement in coastal engineering in Britain, Hooke (1999) suggested that engineering geomorphology was in a second phase of answering geomorphological questions, providing geomorphological information and implementing management, so that a third phase of major development in the future could be envisaged, involving modelling and predicting responses in ways that adequately deal with complexity, positive feedback, non-linearity and holism. She suggested (Hooke, 1999) that geomorphologists arguably have the potential for another major leap forward, stimulated by theoretical and technological developments, in which the results of research feed directly into environmental engineering, provided that the requisite spatial and temporal data are available. A similar prospect was elaborated by Lang (2011), arguing for a more holistic approach to marry process understanding and evolutionary information – 'dreaming of a computer modelling framework, an "Earth surface simulator", to provide a unifying platform: something like GCM technology, representing dynamic process interactions, and including interfaces to the lithosphere, biosphere and atmosphere'. Such attractive prospects are becoming reality in branches of geomorphology (e.g., Gregory et al., 2008), but if geomorphologists do not realize the potential then others who are discovering the need for the geomorphology agenda could adopt it. In relation to coastal management the US Army Corps of Engineers (USACE), the National Oceanic and Atmospheric Administration (NOAA), and the Federal Emergency Management Agency (FEMA) are in the early stages of developing a tri-agency initiative entitled 'SAGE' (Systems Approach to Geomorphic Engineering) to pursue and advance a comprehensive view of shoreline change and to utilize integrated methodologies for coastal landscape transformation to slow/prevent/mitigate/adapt impacts to coastal communities from the consequences of climate change. Their concept is intended to utilize an holistic approach for exploring the idea of hybrid engineering, linking 'soft' ecosystem-based approaches with 'hard' infrastructure approaches to develop innovative techniques and solutions to help in the adaptation of changing coastlines. This addresses the interface of science and public policy and – by incorporating geomorphology in a geomorphic engineering framework – exemplifies the potential that exists. Such approaches need to become embedded throughout geomorphology.

FURTHER READING

Coates, D.R. (1976) *Geomorphic Engineering: Geomorphology and Engineering*. Stroudsburg: Dowden, Hutchinson and Ross, pp. 3–21.

Cooke, R.U. and Doornkamp, J.,C. (1990) *Geomorphology and Environmental Management: A New Introduction*. Oxford: Clarendon Press.

Downs, P. W. and Booth, D.B. (2011) Geomorphology in environmental management. In K.J. Gregory and A.S. Goudie (eds), *The SAGE Handbook of Geomorphology*. London: Sage. pp. 78–104.

Fookes, P. G., Lee, E.M. and Milligan, G. (eds) (2005) *Geomorphology for Engineers*. Dunbeath, Caithness: Whittles.

SAGE: Systems Approach to Geomorphologic Engineering Innovative Approach to Coastal Landscape Transformation www.asbpa.org/conferences/2011abstracts/ChesnuttC.doc

See also Table 18.1.

TOPICS

1. Construct a SWOT analysis to explore the value of geomorphic engineering as a concept.

WEBSITE

For this chapter the accompanying website **study.sagepub.com/ gregoryandlewin** includes Tables 18.1, 18.3; and useful articles in *Progress in Physical Geography*. References for this chapter are included in the reference list on the website.

19

PREDICTION AND DESIGN

The logical culmination of geomorphology/landform science is towards prediction and design, and for both practical and scientific reasons prediction capability is the essential benchmark of scientific quality. Although a 'forecast' is sometimes restricted to prediction in time, 'prediction' is used more generally. But for a variety of reasons forecasting or prediction in geomorphology is difficult and only partially possible in limited circumstances. This chapter first explores conceptually why this is so, and the technical matters that are important where forecasts, hind-casts (where relationships are tested in relation to historically-framed data sets) or novel field, numerical and laboratory 'experimental futures' can be made. Design has been reinvigorated by a 'design with nature' philosophy, embraced by landscape architecture and by ecological engineering, and more recently by geomorphology. Scientific prediction, working with nature, and restoration approaches lead to consideration of how the principle of uncertainty should be applied to pragmatic modelling in general.

Prediction and design were identified as a fourth component for applications of geomorphology in the previous chapter on geomorphic engineering (Table 18.2). This stems from the notion that applicable research, or research results that answer specific management questions, often do not provide the complete answer, because it is also necessary to predict altered future circumstances involving the design aspects of environment. This chapter therefore considers a conceptual framework for prediction (19.1), followed by awareness of design in geomorphology (19.2), progressing to achievements (19.3), and further potential conditioned by uncertainty (19.4).

19.1 A conceptual framework

A prediction is often thought of as a statement relating to the future but in geomorphology is used not only to suggest future occurrences but also to determine spatial distributions or processes in areas for which we do not have geomorphic data. Predictions from models now benefit from effective methods (Wilcock and Iverson, 2003) but scope remains for reducing their complexity (Murray, 2003) as well as making them realistic and reliable.

Future prediction embraces forecasting, and two particular groups of conceptual problem need to be considered when using forecasting tools. Forecasts are always liable to be wrong, despite some analytical outcome producing impressively definite values. *Accuracy* represents the closeness of the forecast to the targeted phenomenon, *precision* the exactness in values achieved. A forecast can be very precise but way out, or it can be in the right zone but only that. Generating forecasts to several decimal places is no great help if they are an order of magnitude out, whereas an approximate figure may be quite useful. Predictions of threshold or catastrophic changes of state, or form metamorphoses, may be particularly valuable, but it is important statistically to define the reliability of predictions wherever possible (their probability levels, or variances). In graphical representations, this means plotting % confidence limit bands on linear graphs, or producing data spreads in the form of 'box and whisker' plots. In an array of data values, an enclosing box contains 50% of the data and the line 'whisker' covers the whole range. The conceptual point to be made is that reliability and possible error needs defining in some way.

Geomorphologists have an entire toolkit of methods for prediction and forecasting on the basis of process controls (C factors) and many are direct imports from hydraulic and geotechnical engineering. They include numerical and physical modelling, heuristic relationships that have been established empirically, and relationships authenticated for specific landforms and processes – meander development, hillslope evolution, or sediment transport. Some predictions are of un-scaled development, general sequences as to how a slope will evolve but not when and why for example, while others are couched in probability terms as in the likelihood of extreme events. But three limitations need to be recognized here. The first is that the 'boundary conditions' for process activity are changing (B factors), not least because of the unfolding history of human activity (A factors). In a sense, in truly evolutionary systems, what happened last becomes the boundary condition for what is to happen next. Predictions are difficult to make if historical contingencies and complexity are involved at any particular site, and especially if climate change and human engineering are in prospect. A second limitation lies in the highly varied nature and reliability of relationships that have been established: some are empirical while others are based on surrogates for universal laws and often represent simple form abstractions like channel width or mean slope angle. These work reasonably well, but they may not incorporate key factors that are difficult to measure or obtain data for, such as riverbank strength when predicting river channel changes. Concepts including error, sensitivity, representativeness and validity are involved, as discussed in Chapter 14.

An **ABC** framework may nevertheless be used by environmental professionals to design, develop and evaluate interventions leading to large-scale behavioural change; this may help conceptually in explaining what has to be involved in making geomorphological predictions. As outlined in Chapter 16, the greatest morphological changes on the Earth's surface, both qualitatively and quantitatively, are now those wrought by human agency (**A**). They include the tallest building (currently the Burj Khalifa in Dubai at 828m) and the largest open pit mine (Hibbing, Minnesota, at 180m deep). It has been estimated that the volume of Earth movement through human agency now exceeds that by natural processes, so that forecasting is not in the hands of geomorphologists but rather in those of architects and engineers, and indeed in the hands of planners themselves. These changes are somewhat unpredictable: the appropriate forecasting device here may be scenario planning using intuitive thinking and professional expertise (e.g., Wright and Cairns, 2011) to which geomorphologists with a number of skills (Downs and Booth, 2011: Table 5.4) may contribute as part of a team (**Table 19.1**). In getting to a range of forecast outcomes ('best case, 'worst case', etc.), it may be possible only to predict some elements with any certainty, and without quantitative definition. Agent-based modelling (ABM) may also be used to allow decision making to be incorporated into models that include human-environmental interactions, to overcome the limitations of existing, highly empirical approaches (Wainwright and Millington, 2010).

In an anthropogenically-transformed world, rain continues to fall and rivers to flow, so that geomorphological processes (**C** factors) proceed in a modified form alongside human structures (e.g., Vörösmarty et al., 2003). These may be called GM processes, with GM (genetically modified) usually being thought of as a biological expression involving gene manipulation. However 'genesis' is a much older word, used for the first chapter of the Bible, and very broadly for 'genetic' classifications. To envision geomorphological processes as now being 'genetically modified' usefully directs attention, and in particular it enhances the need to study what happens in extreme events (Chapter 17). Many structures are designed for a century-scale lifespan and use environmental data during their design, lasting for the same time period or even less. Citizens also use their decade-long or lifetime experience to frame their expectations, so they may need knowledge of what could happen in rare events beyond this, especially in a changing climatic environment of which no one has experience, however 'controlled' the environment may appear to be. Engineering protection and engineering structures can fail when confronted with conditions beyond design expectations, whilst in the natural or artificial spaces between such structures environmental processes continue to operate. Geomorphology is still in action. Knowledge may be gained through comparing modified with unmodified 'benchmark' environments, in 'before and after' studies, and by specifically examining processes where human structures are involved, as in flood embankment-breaches or inter-bank sedimentation rates, and where soil erosion is active (Montgomery, 2007). Application fields are given in **Table 19.2**.

Physical processes are also affected by *boundary conditions* (B). Systems have structures, are interconnected, and behave in particular ways, but the external environment within which they are set also affects them. In open systems this may involve the exchange of matter and energy (Chapter 2), including variable inputs from climates, tectonic regimes and biological factors. For example, biological factors have influenced morphology for a very long time as well as continuing to do so in a modified world. Figure 19.1 plots the relationships in the geological record between channel types recorded and the development of plants on Earth. An organic-free Earth would be different in many respects, without limestones that were formerly coral reefs, fossil fuel deposits, and with only regolith (broken-up rock) rather than soil, for example. At the present day, 'soft engineering' (Chapter 18) also makes use of the fact that vegetation has physical and chemical effects on environments: this includes mitigating pollutant transfers and in part determining riverbank stability and sedimentation (Gurnell, 2013).

Finally, there are the *process controls* (C) that, within bounded systems, directly determine physical form responses to the forces applied. Here there are a limited number of established force-form relationships, some characteristic-form or process rate outcomes for geomorphological domains, and the modelling of forms through notional, numerical and physical experiments. These involve what Huggett (1980) called the parsing and modelling phases (see Chapter 2 and Section 5.1). Investigative techniques are discussed for many geomorphological areas in Goudie (2005), but it has to be appreciated that whilst many studies are now technically sophisticated and quantitative their outcomes do not often lead to predictive models. Those that have done include: relationships in dimensionally-balanced mathematical equations; statistical correlations, including empirical multivariate relationships; heuristic formulae, where something is formally defined in terms of a set of factors, as in F= (P, M) *dt* (Chapter 5); scatter plots, where characterized samples are plotted on a two-variable graph or three-dimensional diagram in order to reveal data sets in one or more parts of the plot field, separable

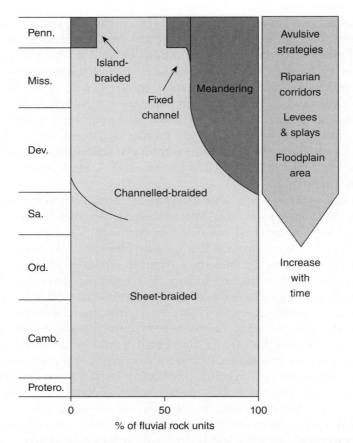

Figure 19.1 River channel types as changing with Palaeozoic plant evolution (after Gibling and Davies, 2012)

by a discrimination line or threshold surface which may be mathematically defined; and two- and three-dimensional model outcomes which need to be validated as applying to their real-world counterparts by some means or other, either visual similarity or, more satisfactorily, real-world/model comparison using form metrics. Clearly there is now an enormous range of analytical tools, numerical and visual, available for geomorphological analysis, and these are greatly helped by computer technology (Wilcock and Iverson, 2003). The fact is that some models are deterministic, resulting from universally valid relationships, whilst others are heuristic, essentially estimating relationships empirically. The second group work because they have been trialled, but they may be of limited reliability where applied outside the conditions in which the proving trials took place. Finally, relationships are usually derived for what might be called the easy bits of landscape, namely simple measures of slope, length or volume. As we have seen, scaling up to emergent and complex forms is logically and in practice difficult, and for the most part the landscape is taken to pieces for analysis.

Design of landscape is closely allied with prediction, in turn reliant upon modelling. Models now available are increasingly sophisticated so that prediction is greatly enhanced (Wilcock and Iverson, 2003), with the uncertainty limitations on predictive modelling (Haff, 1996) being addressed so that the construction of successful predictive models for large geomorphic systems can be attempted. In his chapter in the volume on *Prediction in Geomorphology* Haff (2003) includes design, drawing attention to alternative futures by suggesting that this is likely to become an increasingly important tool, and noting that the corresponding methodology is being used to analyse future climate trends (IPCC, 2001; Schneider, 2001). This raises the issue of how the best science can be available to, and used by, decision makers (Mills, et al., 2008) – a dilemma reminiscent of the paradigm lock (**Figure 18.1**). In considering design, landscape 'alternative futures' recognize the available knowledge base, and require not only use of the best available science and comprehending landscape managers, but also incorporation of the views of stakeholders. This is not an easy task but is one that is being addressed by ecologists (e.g., for the Willamette River Basin, Oregon, see Baker et al., 2004), prompting the question of whether geomorphologists should also be engaging more widely in alternative futures analysis.

Finally, there is not much sign of some grand 'theory of everything' for geomorphology; instead there has been the piecemeal accumulation of useful tools to add to the kit, and some that have been thrown out as well. When an open-eyed and theory-laden geomorphologist stands before a new rolling landscape, there are some things he or she can explain and predict, and others that cannot be, or at least not yet.

19.2 Design in geomorphology

Whereas prediction is the estimation of a condition in another environment or in the future, design involves devising an environmental plan. Design can be simply defined as a mental plan or scheme of attack, but to guide practice and education in the information systems and software design domain Ralph and Wand (2009) clarified the understanding and usage of design and related concepts (**Figure 19.2**). Design of structures has been integral to engineering, with results now characterized as hard engineering in contrast to soft engineering that seeks ways of design and on-going maintenance that are as 'natural' as possible.

As geomorphology concentrates on the Earth's surface and its landforms, including the 'natural' inherited design but embracing ways in which human impact has already been sustained, the discipline should be open to design of the Earth's surface for future scenarios. With more intensive investigations of processes, study of the impacts of human activity, and recognition of the effects of recent evolutionary changes, it is arguably a logical extension of previous research to become concerned with landscape design (e.g., Gregory, 2004a), with the further suggestion that training should include a design project in just the same way that the design project has become a feature of the training of engineers (Gregory, 1992: 333). The suggestion was progressed for fluvial geomorphology where it was put forward (Newson, 1995: 420) that the basis for such environmental design is 'a set of clear prescriptions which, however, must be developed, presented and approved *in context*'. The context arises because, although not located at the core of physical science, geomorphologists have the advantage of an acute knowledge based on the boundary conditions (the local site terrain and form dynamics) within which physical laws operate, and are able to contribute at a local scale where the political, social and economic goals are clearest. Subsequently the idea of a design science in geomorphology was introduced (Rhoads and Thorn, 1996: 132), arguing that 'the growth of geomorphology as a practical profession requires that geomorphologists continue to devote effort to developing and refining a design science to support this profession. Such a design science can provide the basis for professionalization of the discipline by codifying a body of information, tools and skills for licensing or certification programs'.

That relatively little attention had been given to geomorphic design is surprising in view of a book *Design with Nature*, written by a landscape architect at the University of Pennsylvania, Ian McHarg (1969), who proposed ideas of significance beyond landscape architecture, itself potentially so apposite for landform science. First developed in respect of the city, it was suggested in 1964 (McHarg and Steiner, 1998) that an immeasurable improvement could be ensured in the aspect of nature in the city, in addition to the specific benefits of a planned watershed or drainage system. The book *Design with Nature* provided a method whereby ecology was used to inform the planning process, and the intent and language used appeared subsequently in the 1969 National Environmental Policy Act and in other legislative instruments in the USA.

Early roots of a 'design with nature' approach can be traced to Europe in the mid 19th century (Petts et al., 2000); however, progress was slow (see Gregory, 2000; 2004a; 2006) and it was the disciplines of landscape architecture and then ecology that became most significantly involved. In ecology and ecological engineering, 19 ecological design principles were enumerated (Mitsch and Jorgensen, 2004), prompting speculation that the same could be done for geomorphology. Influential ideas had been developed in relation to rivers where engineers (e.g., Winkley,

1972) had advocated the notion of working with the river rather than against it – often such ideas were a reaction to impacts of channelization or consequences of other management concerns such as cost, maintenance, side effects, and the destruction of habitat.

Reflecting on the way in which *Design with Nature* evolved, McHarg (1996) concluded that although a general ecological foundation was influential, there had been insufficient input from hydrology, geomorphology and branches of physical geography, observing (McHarg, 1996: 91) in retrospect that 'geography and the environmental sciences were conspicuously absent', and that 'geomorphology was the integrative device for physical processes and ecology was the culminating integrator for the biophysical. These contributed to understanding process, meaning and form' (McHarg, 1996: 331). A landscape architect (McHarg, 1996) therefore saw the potential significance of geomorphology before geomorphologists did!

19.3 Achievements

Existing design achievements involve interactions with, and sometimes origins within, other disciplines. Two imperatives, both probably stimulated by experience gained in hard engineering, emerged: the need for softer engineering methods and for restoration.

Hard engineering has been applied to solve specific environmental problems such as flooding or coastal erosion, implemented using man-made structures. As it has been found that such hard approaches can lead to unintended environmental consequences, often in adjacent areas, the alternative soft engineering attempts to avoid harsh man-made structures, use natural materials as much as possible, and utilize existing processes wherever feasible – thus working with nature rather than against it. As such imperatives became more popular, geomorphologists became involved (Gregory et al., 1985) because of the way their research had investigated the consequences of hard engineering projects – thus progressing to consider alternatives that were available through soft engineering. The consequences of river channelization were especially dramatic (Chapter 18). Brookes (1985) summarized hard engineering methods, the existing spatial extent of channelization, and the physical, biological and hydrological consequences. Geomorphological research (e.g., Brookes, 1988) subsequently demonstrated how the consequences of channelization were more extensive than previously appreciated so that alternative approaches including biotechnical engineering, morphological alternatives and instream habitat devices (Brookes, 1985), all within the compass of soft engineering methods, were increasingly attractive. Further progress was achieved by studies documenting the magnitude and character of river channel changes arising from reservoir or dam construction, from catchment land use change, from urban development, from other point changes and from water transfers (Brookes and Gregory, 1988), demonstrating how the significance and extent of such changes were much greater than previously anticipated. Such collated research results, and the research on specific areas, inevitably led to an appreciation of the benefits of soft engineering methods, the need to envisage new methods of managing rivers, and the requirement for an holistic approach. Similar approaches also apply to coasts and Hooke (1999) explained that radical changes in both policies and decision-making frameworks had taken place so that the approach to coastal as well as river management adopted by the British Government was to 'work with nature'. More generally engineering interventions failed to resolve symptoms of coastal problems and even generated new ones so that our understanding of them needs to be within a coastal geomorphology context that emphasizes time and space controls on the interaction of energy and sediments (Orford, 2005). Revealing the consequences of hard engineering projects encouraged the search for ways of mitigating undesirable effects as much as possible. More explicit consideration of the coast as a system and the enhancement of desirable system properties such as resilience (Nicholls and Branson, 1998) is required.

Awareness of the need for design with nature was complemented by greater attention being given to restoration, a theme which has proven to be of wider interest because of the concern for whether nature is in large part a social construct (i.e., do environments exist before the environed creature? – Attfield, 1999), followed by related questions including does nature exist; can nature be reconstituted, recreated or rehabilitated; and how sustainable is restoration? Restoration has been applied to river systems where antecedents date from water quality restoration with subsequent influences (**Table 19.3**), but the value of 'faked nature' can never approach that of natural undisturbed systems whose values are defined, in part, by the longevity of their continuous existence and evolution (Elliott, 1997). Issues include whether restoration is possible physically and whether a restored system can ever have the same value as a natural system (Elliott, 1997). A range of different terms has emerged (e.g., Gregory, 2000: 265; Downs and Gregory, 2004: 240) really depending upon whether it is a matter of making the restored condition look more natural or whether it is appropriate for the particular location (**Table 19.4**). The terms proposed can be grouped into three categories (Gregory, 2002): a general category of restoration (which includes the act of restoring a river to a former or its original condition); a category aiming to produce a more natural condition (including what have been called re-establishment, enhancement, rehabilitation, creation, naturalization, recovery); and full restoration to some prior condition (terms used include recovery, full restoration, reinstatement).

Perspectives on river channel restoration (Brookes and Shields, 1996a) included issues embracing an evaluation of environmental impacts, integrated approaches, large rivers, natural recovery as an option, and the need for a project management procedure (Brookes and Shields, 1996b). Such objectives together with challenges identified include improved basic scientific understanding and the appropriate legislation to facilitate restoration. In concluding with an approach to sustainable river restoration, broad guiding principles were devised (Brookes and Shields, 1996c), introducing a theme that has been subsequently developed (e.g., Downs and Gregory, 2004) as elements of design for river channel landscapes (**Table 19.5**). Reflecting on the progress of restoration, Small and Doyle (2012) argued that because some restoration became dominated by a specific classification approach, such as the one by Rosgen (1994; 1996), river restoration now needs to balance numerical analysis with a more general systems understanding, typical of a classification-based approach. Restoration, as the process of returning a river or watershed to a condition that relaxes human constraints on the development of natural patterns of diversity (Frissell and Ralph, 1998), has now been attempted in a range of areas in many landform environments and this experience has been reviewed (e.g., Kondolf et al., 2007). All of this has to be achieved within a cost-benefit framework so that the costs, both of implementation and maintenance, are minimised by comparison with alternatives when achieving objectives such as human safety and both property and habitat protection.

Restoration is now widely undertaken and can relate to areas affected by dam decommissioning (Graf, 2001; Heinz Center, 2002), to an integrated approach for coastal geomorphology where current restoration research is exploring the way that natural and human processes are being integrated (Jackson et al., 2013), and urban areas where management should be focused upon an holistic basin-based approach (Gregory and Chin, 2002). Specific examples of design with nature are shown in Table 19.6. Lessons learned from landform adjustments and the practice of restoration can be combined to give basic principles: an example for the design of whole river landscapes (**Table 19.5**) in river channel design (Gregory, 2004b; 2006) includes context, implementation, design and post-project considerations.

19.4 Potential and uncertainty?

Some decades ago it was thought it might be possible to make a single prediction or create one design appropriate for a particular location in space or in the future. However, realizing the complexity and multiple outcomes (Chapters 7 and 14) leads to an awareness of the degrees of uncertainty that can prevail. Limitations in relation to prediction were outlined (Section 19.1 above), and the progress achieved in design with nature and with restoration approaches may give the impression of a high degree of confidence associated with landscape plans and strategies. Generally it has

been asked (Murray et al., 2009) whether we can develop 'Earthcasts' analogous to weather forecasts, involving both gradual changes and extreme
landscape-changing events. Paradoxically however, as greater understanding has been achieved, awareness of uncertainty has increased. Recently
recognized in many environmental disciplines, and perhaps influenced by
the uncertainty principle of quantum mechanics, the uncertainty idea was
appositely encapsulated long ago by Francis Bacon in *The Advancement
of Learning* (1605) with the words:

> *If a man will begin with certainties, he shall end in doubt;*
>
> *But if he will be content to begin with doubts, he shall end in
> certainties.*

Introducing the volume on *River Restoration: Managing the Uncertainty
in Restoring Physical Habitat*, Darby and Sear (2008: vii) conclude that
'It is evident that the designers and managers of stream restoration
projects are *inevitably* confronted with uncertainty'. They proceeded
to develop a typology (Wheaton et al., 2008) against the background
of two sources of uncertainty suggested for integrated assessment (van
Asselt, 2000), those associated with variability and with limited knowledge. Five alternative philosophical strategies for dealing with uncertainty range from ignore it, through to eliminate it, reduce it, cope with
it, or embrace it (Wheaton et al., 2008: 32). Some restoration schemes
have been considered to have failed to some extent (e.g., Frissell and
Nawa, 1992; Smith, 1997; Kondolf et al., 2001) sometimes because of
the uncertainty involved. To achieve sustainable river restoration as a

Table 19.6   Examples of geomorphological design with nature

Purpose	Specific Investigation
Design of drainage systems on reclaimed mine areas	www.infomine.com/library/publications/docs/Beersing.pdf
Geomorphic design that could create better valley fills in Appalachia	www.statejournal.com/story/19962780/geomorphic-design-could-create-better-valley-fills-in-appalachia
Agencies such as WET (Water & Earth Technologies) which proclaim Geomorphic and Land stream design	http://wetec.us/geomorphic-design
Environment Agency in the UK in which the role of fluvial geomorphology in fluvial design is explained by Maas and Brookes	http://evidence.environment-agency.gov.uk/FCERM/Libraries/Fluvial_Documents/Fluvial_Design_Guide_-_Chapter_3.sflb.ashx

management goal requires viewing river restoration in its catchment and temporal perspective, leading to a resilience-based definition of sustainable river restoration: 'The ability of the river, as restored, to absorb and respond to intra-annual, inter-annual and other cyclical trends in flow and sediment discharge without undergoing such adverse sediment transport and morphological changes that management objectives, including conservation, water resources, and channel hazard minimization, are compromised' (Gregory and Downs, 2008: 254).

To incorporate uncertainty into design-with-nature is 'learning by doing' (Walters, 1986) and is best addressed by adaptive management, defined as '... an innovative technique that treats management programs as experiments'. Rather than assuming that we understand the system we are attempting to manage, adaptive management allows management to proceed in the face of uncertainty (Halbert and Lee, 1991: 138), marking a distinct break from river channel management based almost entirely on experience and intuition (Hemphill and Bramley, 1989) that has tended to result in conservative solutions and an unwarranted belief in design 'certainty'. This highlights the importance of post-project appraisal in determining success, and the need to adjust programmes and policies in the light of experiences gained during the adaptive management programme.

Some aspects of uncertainty can arise from the resilience (Chapter 15) of a system, the term being first used in population ecology to characterize the magnitude of population perturbations that a system could tolerate before changing into some qualitatively different dynamic state (Holling, 1973; May, 1977), but more recently used (Hughes et al., 2008: 93) to describe the ability of an ecosystem to regain a functional state following a disturbance, with the rapidity of this process being a measure of its resilience (Waring, 1989). This is similar to one usage of the term sensitivity in geomorphology. The 'uncertainty' challenge is not unique to geomorphology (Downs and Booth, 2011); it applies generally across the environmental sciences and into civil engineering (e.g., Brookes and Shields, 1996b; Johnson and Rinaldi, 1998), implying the use of approaches that are risk tolerant rather than risk averse (Clark, 2002) and that maximize the chance of beneficial discoveries (McLain and Lee, 1996) as a function of 'learning by doing' (Haney and Power, 1996).

Uncertainty features in geomorphological modelling (Odoni and Lane, 2011) because geomorphological models produce understanding that is provisional because of uncertainties and unknowns that we cannot avoid, with parameter uncertainties focused upon the mental images that the geomorphological modeller or community brings to bear upon the models they develop. Geomorphology has always studied landforms as 'nature's design' so that, as drivers of change have increased, opportunities have been provided to ascertain what has changed and why. The concomitant opportunities to suggest remedial solutions, often by restoration, are being explored, but an appreciation of uncertainty is required to visualize the 'bigger picture'. Progress towards that bigger picture inevitably requires predictions conditioned by uncertainty, and should lead to design attempts constructed to allow for adaptation to be embraced.

FURTHER READING

Darby, S. and Sear, D. (eds) (2008) *River Restoration: Managing the Uncertainty in Restoring Physical Habitat*. Chichester: Wiley.

Gregory, K.J. (2004) Human activity transforming and designing river landscapes: a review perspective, *Geographica Polonica*, 77: 5–20.

Gregory, K.J. and Goudie, A.S. (eds) (2011) *The SAGE Handbook of Geomorphology*. London: Sage. (See especially Chapters 3, 4, 7, 9 and 10.)

McHarg, I.L. (1969) *Design with Nature*. New York: Natural History Press (subsequently issued as McHarg, I.L. (1992) Design with Nature). Chichester: Wiley.

McHarg, I.L. (1996) *A Quest for Life: An Autobiography*. New York: Wiley.

Rhoads, B.L. and Thorn, C.E. (1996) Towards a philosophy of geomorphology. In B.L. Rhoads and C.E. Thorn (eds), *The Scientific Nature of Geomorphology*. Chichester: Wiley. pp. 115–43.

Wilcock, P.R. and Iverson, R.M. (2003) *Prediction in geomorphology*, Geophysical Monograph, 135. Washington, DC: American Geophysical Union.

TOPICS

1. Should environmental design be an integral part of geomorphology higher education courses?

2. The geomorphological consequences of global change are mentioned at the end of Chapter 18. Should these be a major focus for geomorphologists and what more should be done?

 WEBSITE

For this chapter the accompanying website **study.sagepub.com/ gregoryandlewin** includes Figure 19.2; Tables 19.1, 19.2, 19.3, 19.4, 19.5; and useful articles in *Progress in Physical Geography*. References for this chapter are included in the reference list on the website.

CONCLUSION

CONCLUSION

20
THE CONCEPT OF GEOMORPHOLOGY

What sort of a science is geomorphology itself? It is an Earth surface sci-
ence, like some others, and one that interacts with parent sciences (phys-
ics, chemistry, biology, social sciences) and neighbouring sciences (geology,
hydrology, glaciology, atmospheric sciences and soil science). Organized as
an Earth Science or a branch of Geography, it has also begun to develop
specialisms and has subdivided to produce others (e.g., speleology). It also
retains distinctive characteristics, involves a fascinating form of human
enquiry, has significant practical social value, and is recognizing its own
conceptual merits and difficulties. The current state of geomorphology
within the range of the sciences is reviewed, using a SWOT approach.

Geomorphology, defined in Chapter 1 as the science concerned with the
forms and processes on the Earth's land surface, is an Earth surface science generally dealing with timescales of seconds (s, the SI base unit) to
millions of years (10^6 yrs. or s^{13}). Spatial scales usually cover a range from
m^{-3} (mm) to m^5 (1000 km). Some analyses exceed these bounds, for example laboratory work involving micrometre-scale particles (10^{-6} m or μm).
There are other surface sciences that deal with the interface between solids, gases and liquids at different scales: generally at a smaller operating
scale, these are concerned with such matters as friction, wear, corrosion,
lubrication, diffusion, osmosis, plating, surface cleaning and cosmetics.
Boundary states and processes are very important to all material sciences,
as in hydraulics or the atmospheric sciences, whilst the study of interacting
surfaces in motion now has a name of its own (tribology).

At the other end of the scale, moons and other planets also have Earth-
analogous surface forms and their study may or may not be included in
geomorphology (Melosh, 2011; Greeley, 2013). Etymologically, the prefix
ge- may not seem appropriate, and some alternative names have been suggested, such as 'selenomorphology' for studying the surface of the moon.
It has been argued (Baker, 2008) that for geomorphology to be a complete
science of terrestrial landforms and their generative processes, it needs a
planetary context. Users of the ancient term 'geometry' have not restricted

themselves to land measurement, and there seems to be no disqualifying reason that precludes use of 'geomorphology' for other planetary bodies. Martian landforms have fluvial and glacial features of considerable interest which geomorphology, amongst other sciences, is well equipped to understand (Kleinhans, 2010).

Strictly speaking a boundary, like a line in mathematics, may be considered to have no substance. But the Earth's boundary is transitional with an interpenetration of gases, liquids and solids. The surface is mobile, developing by solid extension (differential uplift, volcanism, etc.), solid removal (sculpted by atmospheric, liquid and solid movement) and by accretion (deposition of entrained material solids and precipitates). Analysis of the surface necessarily involves parent science processes (physics, chemistry, biology and the social sciences) and interaction with neighbouring sciences (geology, hydrology, oceanography, glaciology, cryology, atmospheric sciences and soil science) which have their own concerns with near-surface or boundary phenomena. It may make sense to link these together: about half of Benn and Evans's book *Glaciers and Glaciation* (2010) deals with surface forms, whilst texts on Fluvial Geomorphology generally cover elements of hydrology and hydraulics (e.g., Knighton, 1998; Charlton, 2008). So, is a river an Earth form (with width, depth, etc.) or a flow of water? It serves no practical purpose to be too meticulous about such distinctions, except to realize that the objectives of one science may be different from those of another. Thus the hydraulics engineer may see the river outline in terms of flow resistance with the supplementary objective of designing for form stability, the hydrologist may see quality and yield in a liquid phase of water movement, and the geomorphologist the dynamic containing form. It helps to work together, and there is a need to have 'a dialogue about the nature of mutually valued questions and to use mutually-acceptable methods' (Rice et al., 2010). Furthermore, many human problems depend on an interaction between socially generated activities and physical processes: the drivers of change are combinations that require cross-disciplinary collaboration and understanding.

So what is distinctive about geomorphology? It does not have ready-made units of study like biology that, at least at first sight, has individual entities like plants and animals. These constitute the stuff of classification, whether in a rank-based grouping hierarchy (domains, phyla, classes, orders, families, genera and species), in ecosystems as entities (genes, cells, tissues, organs, organisms, species, populations, communities, biomes and the biosphere), or through cladistics (classification by ancestry, also known as evolutionary taxonomy and phylogenetic systematics). Nor does geomorphology search for fundamental particles and matter, as do physics and chemistry. These sciences may seek to identify them now in the subatomic realm, to classify them as in the periodic table, or search for

Table 20.1 A contemporary conceptual set for systems analysis

Topic	Example	Realizations
Morphometry	River, slope and valley form definition	General parameters or specific geometries
Initiation	Network formation	Geomorphology's 'starburst' phase; network modelling, fractals
Characteristic forms	Slope profiles, channel/valley geometries, glacial and coastal depositional forms, climatic variety, ecological relationships	Dynamic or steady-state equilibrium definition and relationships; self organization and scale similarities/contrasts
Input-output modelling	Sediment cascades	Sediment budgets, including stores and throughput; feedbacks; non-linear relationships
Recycling	Formation-preservation sets	Alluvial landform assemblages, continent-scale patchworks
Trajectories	Hypsometric trends, relaxation forms	Relief change models, threshold definitions
End-states	Entropy maximization	Planation forms
Management	Interdisciplinary scenario planning, anthropogenic impacts	Hazard and degradation resilience, restoration, sustainable design

them in hypothesized form, as in the case of dark matter and energy that are currently believed to constitute around 27% and 68% respectively of the universe. In geomorphology, despite the fact that certain landforms such as the ox bow lake are extremely familiar, they are a heterogeneous though diagnostic collection and not generally set as fundamental units within a sequence of universal landforms.

Geomorphology has identified and classified forms analytically and sometimes arbitrarily, whether as forms like moraines or as measured elements like characteristic slope angles. Named identities are frequently historical and may relate back to pre-science days. The basis of understanding often involves empirical relationships between form element measurements, and between them and physical forces leading to movement, whilst surface quantification may involve the derivation of summary parameters or mathematical approximations. As the timescales over which emergent forms develop usually exceed direct observation possibilities, modelling of the unseen, whether qualitative or quantitative, has been a central occupation (Anderson, 1988). The way the Earth's surface develops has to be formally or informally imagined, a task which students sometimes find difficult in the field when told of a land surface that was once above their heads many thousands of years ago, and this difficulty persists despite the aesthetically appealing and literally down-to-earth nature and human scale of the phenomena studied.

At its study scale for the surface of the Earth, geomorphology necessarily involves spatial heterogeneity, complexity, diachronism and episodicity. There are equilibrium and non-equilibrium states, linear and non-linear processes, and cyclical rhythms and directional trajectories. We have seen that the discipline has been much concerned with 'big picture' concepts like systems (Chapter 2), uniformitarianism (Chapter 3), equilibrium states (Chapter 6), complexity (Chapter 7), and cycles (Chapter 8). In this sense it has not been unlike the early history of physics, dominated by Newton and the unifying concept of gravity. But geomorphology now, like physics, appreciates that what operates at different scales may need to be approached differently, and that synthesis and a whole range of analytical understandings are required (Slaymaker, 2009). A modern-day equivalent of Davis's classic cycle framework for fluvial landform development, and therefore study, might look something like Table 20.1, with concepts like the cycle and equilibrium each having their place, but also now including others like magnitude-frequency and preservability. At a research operational level, specialisms such as modelling, laboratory and dating techniques, design and management occupy particularly valuable niches, with an overall need for an harmonious integration rather than a conceptual monotone. Support is required from one of the oldest traditions of the discipline, that of fieldwork, because the ultimate arbiter of concept value, whether in validating models or discovering new entities, is out there as the Earth's surface itself (Church, 2013). 'Remote fieldwork', even on Mars, is also now possible using electronic sensors and other recently available devices.

In Chapter 1 the development of geomorphology as a science was explained. The discipline itself is a concept and, with hindsight, had geomorphology not been invented what would have been conceived in its place? Throughout the preceding chapters multiple disciplines have been mentioned – could any of these have been alternatives or could they progress only because geomorphology already existed? There are also numerous cases where the expected tasks of geomorphology are being taken over by other disciplines – so does this expose an inherent weakness of geomorphology? The presently conceived state of geomorphology can usefully be assessed using a SWOT matrix approach. The discipline is not a business enterprise, of course, but geomorphology at present does have strengths, weaknesses, opportunities and threats. Table 20.2 summarizes a selection, and having developed opinions from reading previous chapters, readers may make changes, suggest other inclusions, or indeed wish to join in and work to moderate some of those on the right-hand side of the table!

Table 20.2 *The State of Geomorphology:* a preliminary SWOT analysis (book chapters where issues were discussed are indicated in brackets)

Strengths	Weaknesses
Is a visible and tangible everyday outdoor human encounter, with a strong empirical science base	A heterogeneous science, requiring very diverse expertise; fissiparous
The discipline that focuses on the Earth's surface (1, 4)	Potential students can be deterred by science knowledge required
Provides a holistic understanding (2, 3) of the Earth's surface which includes the cultural (16) as well as the physical environment	Phenomena may be equivocally defined
	Relationships and hierarchies yet to be comprehensively and quantitatively modelled (4)
Deals in practical problems like erosion, sedimentation, environmental pollution and hazards (5, 6, 17)	Some environments researched more than others (e.g., tropical, urban)
Extends understanding of threats to human welfare, as in extreme events (17)	Empirical traditions need further theoretical (and mathematically specified) support (6, 7, 13)
Imaginatively lengthens the timespan of human experience (12–15)	Fieldwork-demanding and thus logistically difficult and expensive
Positioned to lead multidisciplinary teams (18)	Slow to develop as an accredited professional group
Can engage with and be appreciated and supported by a reasonably educated and informed public	Not self-marketing effectively enough, and often not understood by general public

Opportunities	Threats
Global climatic/environmental change requires the knowledge extensions the discipline can provide (16, 18)	Not a strong recipient of science funding
Developing new techniques (e.g., in computation, GIS, remote sensing and rapid chemical assay) that provide new analytical openings (1)	No large cohort of researchers, or a tradition of team research
	Requires considerable equipment investment
Needed, with engineering and ecology partners, to devise sustainable environmental design (19)	Selective take-over by other science groups richer in resources and expertise
Media channels increasingly visual, demanding of new material, and geomorphology could be more prominent	Subdivision with denudation of parts of discipline as new subjects developed
Provide understanding of Earth's surface for informed public	Side-lined in urban societies living (perhaps recklessly) in artificially encapsulated environments

As a science area in the modern world, geomorphology has to compete for attention and funding, and for the students who will perpetuate an interest in what it offers. It must also give more attention to problems that are of practical importance for human wellbeing, and there is a need to demonstrate its effectiveness for doing so. The concepts utilized are numerous but are they clearly valuable or shaky, how can they be developed further, and which are likely to survive? We should not regard disciplines, or indeed any concepts, as immutable, and indeed reporting results and promoting ideas have been treated as storytelling (Phillips, 2012b) – we need to know how the 'plot' developed to arrive at the contemporary state of geomorphology.

FURTHER READING

Church, M. (2013) Refocusing geomorphology: fieldwork in four acts, *Geomorphology*, 200: 184–192. Doi: 10.1016/j.geomorph.2013.01.014

Kleinhans, M.G. (2010) A tale of two planets: geomorphology applied to Mars' surface, fluvio-deltaic processes and landforms, *Earth Surface Processes and Landforms*, 35: 102–117. Doi: 10.1002/esp.1895

Phillips, J. (2012) Storytelling in the Earth sciences: the eight basic plots, *Earth Science Reviews*, 115: 153–62.

Slaymaker, O. (2009) The future of geomorphology, *Geography Compass*, 3: 329–49. Doi: 10.1111/j.1749-8198.2008.00178.x

TOPICS

1. Develop and amend the SWOT analysis (Table 20.2).

 WEBSITE

References for this chapter are included in the reference list on the website.

INDEX

Major items in the book text (not the supporting web site material) are indexed, locations and individuals are not included. Page numbers in **bold** refer to tables those in *italics* to figures.

environmental management 141
episodism 27
equilibrium 16, 56–64, 165
 equilibrium-line altitudes (ELAs) 59
 equilibrium shorelines 59
 equilibrium state 28
ergodic reasoning 142
erosion 85
 erosion intensity 105
 erosion rate 114, 115
 erosion surface/planation surface 43
 erosion surfaces 163
erratics 21
essentialism 42
evolution 4, 48, 51
 evolution theory of 23
exogenetic process regimes 125
exogenetic processes 85, 93
exposure ages 115
extrazonal phenomena 49
extremal hypotheses 108
extreme events 105, 184, 185, 206
extrinsic response 67

factor of safety 98, 99, 116
faked nature 212
feedback 73
 feedback loops 153
field studies 153
fissiparist or reductionist trend 52
flood hazard 189
floodplain patchiness 164
floodplains 164
flume studies 57
fluvial geomorphology 200
fluvial hazards 190
fluvial processes 57
fluxes 79, 91
force 93, 95, 96
forcing 90
forcing functions 137–145
forecast 205
form sequences in time *148*, 150
formation time 160
formative events 104, 105, 133
fractals 73
frames 3
frequency/magnitude concepts 133

Gaia theory 17, 81, 178
General systems theory 13, 14, *56*
geoecology 172
geographical information systems
 (GIS) 34
geohazards 184, 185

geoinformatics 36
geological cycle 80
geological timescale *128*
geomorphic effectiveness 118
geomorphic engineer 193
geomorphic engineering 176,
 193–202
geomorphic hazards 183–191
geomorphic processes 90, 93, 94
geomorphic response 67
geomorphic systems 18, 68
geomorphic thresholds 143
geomorphic transport laws 97, 98
geomorphic work 101–109
geomorphological features *32*
geomorphological mapping 188
geomorphological maps 36, 39, 42
geomorphological models 37, 97, 215
geomorphological processes 96, 206
geomorphological techniques **6,7**
geomorphology 1, 2, 4, 5, 8, 31, 37, 46,
 51, 58, 113, 153, 171, 209, 210, 211,
 215, 219, 221, 222, 224
 geomorphology applied 199
 geomorphology branches of 51
 geomorphology coastal 212
 geomorphology engineering 201
 geomorphology environmental 195
 geomorphology fluvial 200
 geomorphology macroscale 52, 68
 geomorphology volcanic 188
geomorphometric map 36, 37
geomorphometry 35, 36
geomorphon 35, 37
geosphere 186
geothermal energy 92
geothermal gradient 92
glacial drainage channels **43**
glacial landforms 163–4
glacial-interglacial cycle 139
glaciation 82
glacier power 107
glaciers 107
global change 196
global climate change 175, 191
global cycles 79–82
global environmental change 1
Global Positioning Systems (GPS) 35
global warming 190
GM processes 206
grade 57, 58, 115
graded profile 56
graded slope 59
graded stream 58
graded time 131